D0609366

ELECTROSTATIC DISCHARGE AND ELECTRONIC EQUIPMENT

A PRACTICAL GUIDE FOR DESIGNING TO PREVENT ESD PROBLEMS

OTHER IEEE PRESS BOOKS

Instrumentation and Techniques for Radio Astronomy, *Edited by P. F. Goldsmith*
Network Interconnection and Protocol Conversion, *Edited by P. E. Green, Jr.*
VLSI Signal Processing, III, *Edited by R. W. Brodersen and H. S. Moscovitz*
Principles of Expert Systems, *Edited by A. Gupta and B. E. Prasad*
Microcomputer-Based Expert Systems, *Edited by A. Gupta and B. E. Prasad*
High Voltage Integrated Circuits, *Edited by B. J. Baliga*
Microwave Digital Radio, *Edited by L. J. Greenstein and M. Shafi*
Oliver Heaviside: Sage in Solitude, *By P. J. Nahin*
Radar Applications, *Edited by M. I. Skolnik*
Principles of Computerized Tomographic Imaging, *By A. C. Kak and M. Slaney*
Spaceborne Radar Remote Sensing: Applications and Techniques, *By C. Elachi*
Engineering Excellence, *Edited by D. Christiansen*
Planar Transmission Line Structures, *Edited by T. Itoh*
Introduction to the Theory of Random Signals and Noise, *By W. B. Davenport, Jr. and W. L. Root*
Teaching Engineering, *Edited by M. S. Gupta*
Robust Control, *Edited by P. Dorato*
Writing Reports to Get Results: Guidelines for the Computer Age, *By R. S. Blicq*
Multi-Microprocessors, *Edited by A. Gupta*
Advanced Microprocessors, II, *Edited by A. Gupta*
Adaptive Signal Processing, *Edited by L. H. Sibul*
System Design for Human Interaction, *Edited by A. P. Sage*
Microcomputer Control of Power Electronics and Drives, *Edited by B. K. Bose*
Advances in Local Area Networks, *Edited by K. Kümmerle, J. O. Limb, and F. A. Tobagi*
Load Management, *Edited by S. Talukdar and C. W. Gellings*
Computers and Manufacturing Productivity, *Edited by R. K. Jurgen*
Being the Boss, *By L. K. Lineback*

A complete listing of IEEE PRESS books is available upon request.

ELECTROSTATIC DISCHARGE AND ELECTRONIC EQUIPMENT

A PRACTICAL GUIDE FOR DESIGNING TO PREVENT ESD PROBLEMS

Warren Boxleitner
Director of Engineering
KeyTek Instrument Corporation

Published under the sponsorship of the
IEEE Electromagnetic Compatibility Society.

The Institute of Electrical and Electronics Engineers, Inc., New York

This book is set in Times Roman

PRINTED IN THE UNITED STATES OF AMERICA

IEEE Order Number: PC0235-2

Library of Congress Cataloging-in-Publication Data

Boxleitner, Warren.
Electrostatic discharge and electronic equipment: a practical guide for designing to prevent ESD problems/Warren Boxleitner.
p. cm.
"Published under the sponsorship of the IEEE Electromagnetic Compatibility Society."
Includes index.
ISBN 0-87942-244-0
1. Electronic apparatus and appliances—Protection. 2. Electric discharges. 3. Electrostatics. I. Title.
TK7870.B68 1988
621.381—dc19 88-21852

Contents

* Each topic is followed by the guideline roman numeral and page number

* Each topic is followed by the guideline roman numeral and page number

* Each topic is followed by the guideline roman numeral and page number

Preface

Since electronic equipment was first developed, static electricity has been a source of problems for users and designers. In the last few years, however, electrostatic discharge (ESD) has become a source of major problems. This has occurred because newer electronic devices, such as integrated circuits, are much more susceptible to ESD problems than previous devices, such as vacuum tubes. Another trend compounding this ESD susceptibility problem is the spread of sophisticated equipment into home and office environments where ESD is quite common.

Unfortunately, this increase in the occurrence of ESD problems has not been accompanied by a corresponding increase in knowledge about ESD. Today, in the electronics industry, there is probably no other area which is so poorly understood as ESD. Most engineers view the subject of ESD as black magic, and design for prevention of ESD problems as a black art. As a result, a typical engineer takes few conscious design steps to prevent ESD problems. When problems do occur, solutions are randomly applied until the symptoms go away. In many, if not most, cases, the designer doesn't really understand why some solutions fail and others work. Even worse, designers are often not sure whether the problem has really been solved, because they don't understand how to perform accurate verification tests.

Part of this confusion occurs because many engineers attempt to treat ESD as just another form of electromagnetic interference (EMI). While it is true that ESD generates EMI, ESD is not merely EMI. In addition to generating EMI, ESD also directly injects charge into the victim equipment. This direct charge injection results in several special problems that do not occur with normal EMI. In some situations, normal EMI solutions may actually aggravate ESD problems.

This book was written to take the mystery out of ESD. It explains how ESD is generated, and how it affects electronic equipment. This explanation brings ESD out of the realm of black magic and into the sphere of science. Even more important, this book explains how to design equipment to prevent ESD problems. This discussion of ESD design solutions not only includes design guidelines, but explains why they work. It also exposes myths that have developed about ESD and why they are incorrect. Finally, this book discusses the methods of testing for ESD problems. This discussion covers not only the test hardware, but also test procedures and methods that ensure meaningful results. The information contained in the following pages should help the reader prevent many (if not most) ESD problems.

1 A Model of the Electrostatic Discharge (ESD) Event

In order to fully understand electrostatic discharge (ESD) effects, it is necessary to understand the causes. In this chapter, the ESD event is described in detail. To make the explanation more concrete, the description is based on the example of a person walking across a carpet and then touching an item of electronic equipment. The specific item of equipment used in the example is a computer keyboard. However, it should be remembered that ESD can be generated in many other ways, such as movement of paper within a printer, and that ESD can impact any item of equipment or an entire system. The basic concepts discussed will apply no matter how the ESD occurs.

To begin the discussion of the charging process, assume the person is not electrically charged. As the person walks across the carpet, the soles of the person's shoes come into direct frictional contact with the carpet. Depending on the molecular structure of the shoes and the carpet, there is a tendency for one surface to strip electrons from the other. This is commonly called triboelectric charging, although the primary charging effect is probably not triboelectric, but contact charging. Many sources reference a triboelectric series that claims to indicate which dielectric will strip electrons from which. In fact, there has been little success establishing a precise series which can be proven repeatedly by experiment. In one experiment, rayon may strip electrons from rubber, and in the next experiment the opposite may occur. This is thought to be caused by surface impurities on the dielectrics. As a result, it is probably not possible to predict whether the shoe soles will become positively or negatively charged. However, some charge will develop on the shoes and an opposing charge will be left in each footprint on the carpet. As the person walks, the charge on the shoe soles tends to become progressively greater with each step. However, there is a limit on the charge which may be stored.

Opposing the charge buildup is a small return current, some of which flows via the dielectric of the air, but most flows via the shoes and the carpet. High humidity reduces the resistance of most dielectrics and will thus increase the return current. This means the charge on the shoes will reach an equilibrium point where triboelectric charging equals the return current. (Temperature also effects dielectric resistance, but to a much smaller degree than humidity.)

Thus far, the discussion has been confined to the dielectrics of the shoes and carpet. There is also a conductor, which is the human body. Students of electrostatics will recognize that the shoe soles are crude electrets, which have an electrostatic field. As a result of this field, the conductive tissues and moisture layers on the soles of the feet will develop a charge opposed to the charge in the shoe soles. In the process of charging the feet, charge is redistributed within the body. (Except for the skin, most body tissues are fairly good conductors.) This means the remainder of the body, not including the feet, will typically have a charge opposite to that of the feet. (However, the actual situation may be complicated by additional charge sources such as socks and other items of clothing.) As a simple example, assume the shoe soles strip electrons from the carpet. This leaves positively charged footprints on the carpet and a negative charge on the shoe soles. This negative charge on the shoe soles results in a positive opposing charge on the soles of the feet. However, if positive charges have moved to the feet, that leaves the rest of the body with a negative charge. The level of the charge on the human body is limited by the return current previously discussed, which at very high voltage includes corona discharge.

In the previous discussion, only the charging of the person was considered. The example keyboard will now be added to the picture to complete the discussion of the charging process. (A keyboard was chosen for this example because it is an item of equipment commonly touched by the operator; however, all items of equipment within a system must, of course, be designed for ESD immunity.) As the person walks toward the keyboard, assume the triboelectric charging and return currents continue to keep the charge level on the person stable.

During the approach to the keyboard, the charge on the person will generate an opposing charge in the keyboard. Since the keyboard is grounded, its charge will be developed by electrons flowing in the keyboard ground line within the keyboard cable. (Ungrounded items will have their charge redistributed to oppose the charge on the person.) In the example, with the person's body negatively charged, the keyboard will develop a positive opposing charge through loss of electrons, via the ground path to earth. The closer the person approaches, the greater the opposing charge on the keyboard. It should be noted that the rate at which charging current flows in the keyboard ground path is dependent on the speed at which the person approaches. However, even a fast approach would result in a slow rate of rise in charging current. Therefore the charging current that flows prior to the discharge will have no significant impact on the keyboard operation.

More important than the charging current is the actual electrostatic field that exists between the person and the system prior to the discharge. This field can induce unequal voltages within items of equipment. Sufficiently large unequal voltages could even result in destruction of components such as integrated circuits (ICs). (This indicates that the discharge itself is not the only source of potential problems.)

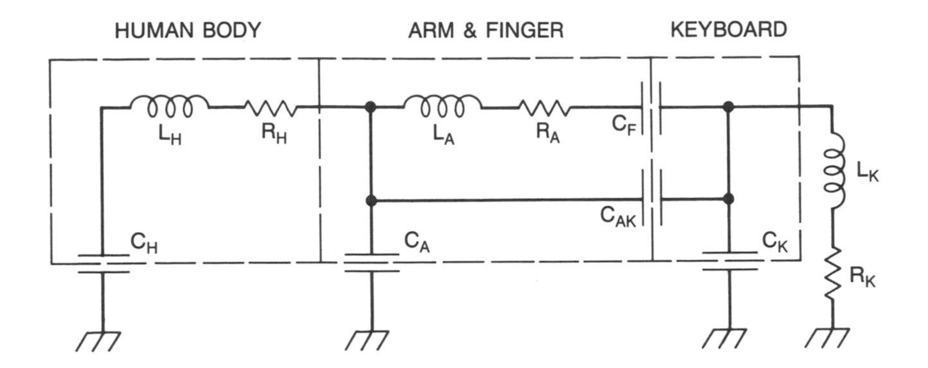

Fig. 1: *Electrical model of human and keyboard ESD system.*

To complete the discussion of the ESD charging process, the circuit in Figure 1 has been developed to serve as a model for the person/keyboard system used in this example. In Figure 1,

C_H = Lumped capacitance between the human body and earth
R_H = Lumped resistance of the human body
L_H = Lumped inductance of the human body
C_A = Lumped capacitance between the person's arm and earth
C_{AK} = Lumped capacitance between the person's arm (and near portions of the body) and the keyboard
R_A = Lumped resistance of the person's arm's discharge path
L_A = Lumped inductance of the person's arm's discharge path
C_F = Capacitance between person's finger, hand, and the keyboard
C_K = Lumped capacitance of the keyboard to earth
R_K = Lumped resistance of the keyboard earth ground path
L_K = Lumped inductance of the keyboard earth ground path.

Resistance and inductance between C_F, C_{AK}, and C_K is very small and thus not included in this model.

Five points should be emphasized about this model:

1. Although lumped values are used here, one should keep in mind that in the real world these effects are distributed. (Transmission line theory would be more suited to precisely describing the ESD process.)
2. C_H, C_A, and C_K are often referred to as "free space" capacitance because the two capacitance elements (e.g., the body and the earth) often have a large physical separation, and may thus approach free space values. It should be noted that this is not always the case. A person physically close to ground will have a higher body capacitance than a person far from ground.
3. There is no inductance or resistance between C_{AK} and the keyboard or between C_A and C_H and earth ground. This means an ESD generator designed to simulate this model must be designed very carefully. Even the inductance of a wire could impact the results dramatically.

4. This model represents equipment which is connected to ground. Hand-held or portable equipment would not have L_K or R_K, but would be the same otherwise.
5. Although this model was developed to explain the person/keyboard example being used, the model is in fact quite general. By changing the R, L, and C values in the model, many other ESD situations may also be modeled. In each case L_H, R_H, and C_H represent the bulk of the charge source; L_A, R_A, and C_A represent portions of the charge source which are physically nearer the discharge point, and through which current from C_H must flow; C_{AK} represents the capacitance from these nearer portions of the charge source, which couples to the electronic equipment; C_F represents the capacitance from that part of the charge source which is closest to the discharge point to the electronic equipment; and C_K, L_K, and R_K represent any item of electronic equipment.

The introduction of the previous model completes the discussion of the charging processes that take place in an ESD event. The remainder of the discussion deals with the discharge phase of the ESD event. In the previous example, the person was poised with their finger almost in contact with a computer keyboard. As the person's finger draws nearer the keyboard, the electrostatic field intensity between finger and keyboard will eventually become so great that dielectric breakdown occurs in the air between them. This begins with a streamer, which establishes an ionized path of conduction, and progresses into the familiar spark, in which the majority of the charge is transferred.

Although the person's approach speed prior to initiation of arc formation is not critical, the speed of approach during arc formation itself is very important. The formation of the arc requires much more time than the duration of the arc. Since the person's finger is moving closer to the keyboard during this long arc formation process, a fast approach will result in a narrower arc gap than a slow approach, even when the voltage level is the same. Therefore, for a fast approach, the voltage is unnaturally high related to the arc gap length. This results in a more intensive discharge with faster associated current rise times and peaks.

A slight modification to the previous model will allow an explanation of the very important discharge process. As seen in Figure 2, the basic model remains unchanged, except that the resistance and inductance of the arc discharge path are added in parallel with C_F. The values for R_S and L_S are not actually constant, but vary during the arcing process. This is especially true of R_S which starts as a relatively high resistance and drops as the air becomes more and more ionized.

Even with its limitations, the model developed gives important insights into the discharge process. When the arc forms, it will first be discharging C_F. The components R_S, L_S, and C_F will form a damped tank circuit. The

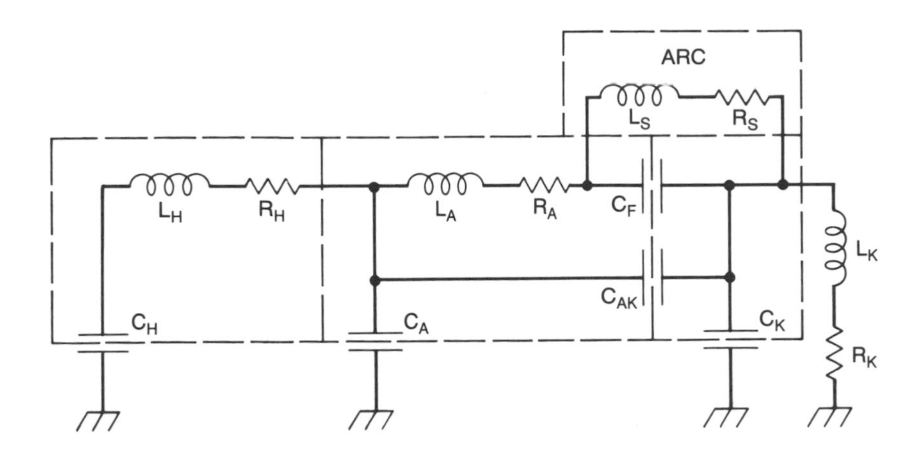

Fig. 2: *ESD model including the arc.*

damping depends on the value of R_S and the tank frequency on C_F and L_S. The value of C_F depends on the size and shape of the person's finger and hand. A smaller hand and more pointed finger will result in a smaller value for C_F, and a smaller C_F would theoretically result in a higher frequency for the tank. However, a more pointed finger will also experience corona discharge at lower voltages. The onset of corona discharge can significantly impact the resulting discharge waveform. In the model, corona could be modeled as a bleeder resistor shunting around C_F, C_{AK}, and C_A. Prior to the actual arc, corona will bleed charge off C_F and even somewhat from C_A and C_{AK}. This means the higher frequency components of the discharge wave will be reduced. Therefore, the maximum frequency of the ESD will depend on the values of R_S, L_S, and C_F, only if corona is not occurring.

As C_F discharges, the parallel combination of C_{AK} with C_K and C_A will also begin discharging. However, the discharge current from this parallel combination must not only go through R_S and L_S, but also R_A and L_A. Also, the capacitance of this parallel combination is larger than C_F. This means the discharge of C_A and C_{AK} will be slower than the discharge of C_F alone. In the case of C_H, the discharge path includes R_H, L_H, R_A, L_A, R_S, and L_S. Also, the discharge path of C_H includes the parallel combination of C_K, with R_K and L_K.

It is important to point out that very little of the discharge currents of C_F, C_{AK}, and C_A flow in the keyboard ground path. Also, any high frequency components of the discharge current from C_H will tend to flow through C_K, not the keyboard ground path. Current in the keyboard ground return is limited to low frequency components of the discharge current from C_H.

The exact value of R, C, and L, in each case, will determine the exact discharge waveform. As previously indicated, the discharge of C_F will tend to create an extremely high frequency. The discharge of C_A and C_{AK} will create high frequencies. Finally, a somewhat lower frequency will be generated by the discharge of C_H. In addition to the frequency ranges indicated above, the fact that the capacitors are discharging will result in a damped oscillation.

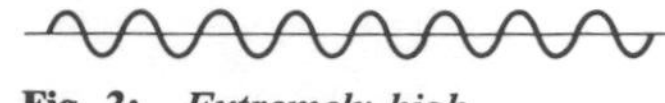
Fig. 3: *Extremely high frequency.*

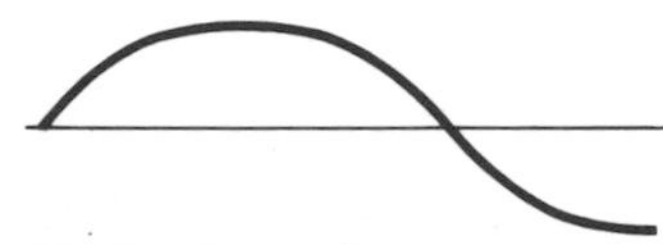
Fig. 4: *High frequency.*

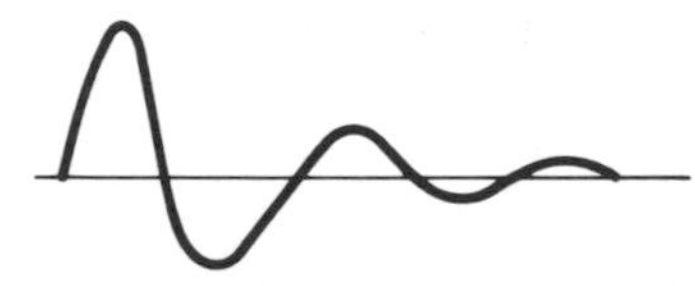
Fig. 5: *Lower frequency.*

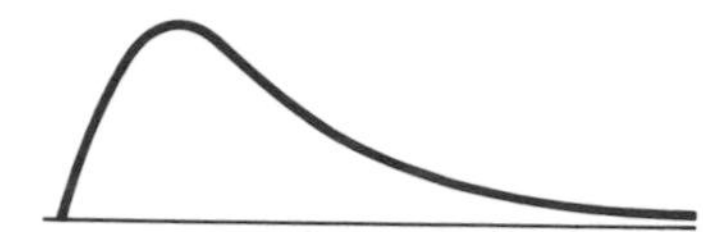
Fig. 6: *Underdamped.*

Fig. 7: *Overdamped.*

This can be either underdamped or overdamped, which is again dependent on the exact R, L, and C values. These various waveform components are indicated in Figures 3 through 7.

Both computer modeling and experimentation indicate that the waveforms will be typically either overdamped or very damped oscillations, depending on the exact position, size, etc., of the person. (Furniture-related discharges may result in underdamped oscillations.) The total current waveform for a typical discharge is shown in Figure 8 (the dotted line shows the waveform with the effect of corona discharge).

In such a waveform the low frequency components will transfer more of the charge than the high frequency components, but the high frequency components will generate the most intense fields. Experimentation has shown that the following time limits can occur for an arc:

T_r (Rise Time) = 200 ps to 70 ns
T_s (Spike Width) = 0.5 ns to 10 ns (if the "spike" exists)
T_t (Total Duration) = 100 ns to 2 μs.

Computer modeling indicates even wider ranges are possible.

In addition to timing differences, peak currents can also vary from one amp to over 200 amps. (Computer modeling indicates even higher, and lower, currents are possible.) With this wide range of responses for different conditions, it is no surprise that the ESD response of electronic equipment often appears unpredictable. Fortunately, statistical methods are available to cope with this problem. The important lesson from this analysis is that relatively high energy and high frequency (5 GHz) signals may be generated by the ESD event. Another important point is that C_F, C_{AK}, C_A, L_A, and R_A have a great impact on the generation of high frequencies. Simple RC models used in the past have ignored these components.

Figure 9 shows a simple RC model that ignores most of the components in a true human model. This circuit carries simplification too far, and thus results in a faulty view of the problem.

This nearly completes the discussion of charging and discharging events, and most analysis ends at this point. However, the story is not quite done yet. More than one experimenter has noted that multiple discharges may occur during a single ESD event. These discharges are at successively lower peak current levels, and are separated by 10 μs up to 200 ms. There are two or three effects which could combine to cause these multiple discharges. Looking again at the model, it is seen that a high value for R_H and L_H would allow C_A and C_F to discharge completely even though C_H is still charged. After C_A and C_F are discharged, the arc would quench. Then C_H would recharge C_A and C_F until the air breakdown voltage is again attained. A new arc would result, and C_A and C_F would again be discharged. This would continue until C_H was fully discharged. Much of the charge, associated with capacitance C_H, is in the soles of the feet, and a majority may even be stored in a layer on the bottom surface of the feet. In this case, R_H would include

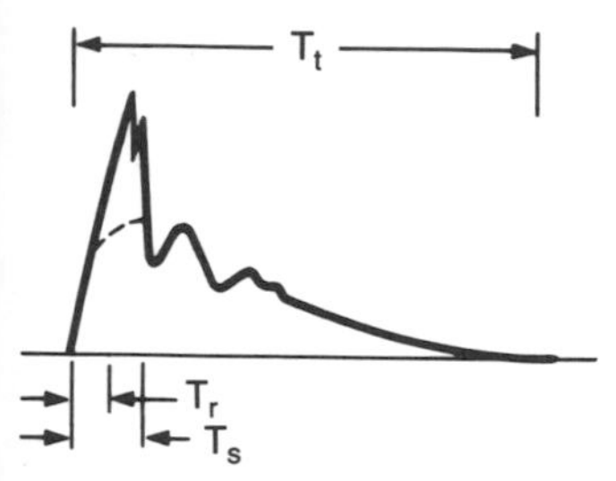

Fig. 8: *Typical human ESD current wave.*

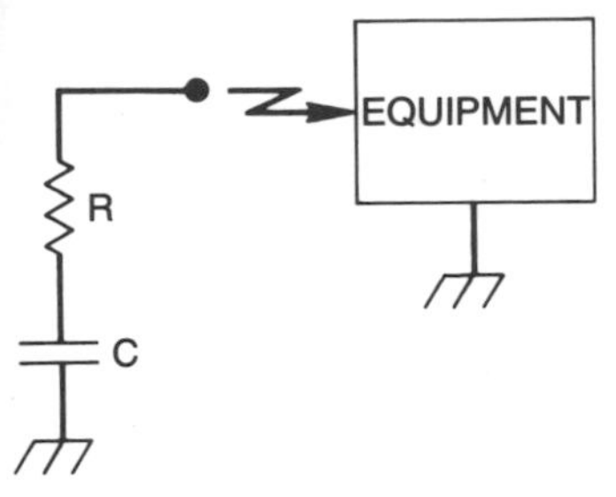

Fig. 9: *Simple RC model of ESD.*

skin resistance which can be relatively large. This would explain discharges separated by tens of microseconds, but it wouldn't explain longer separations. In order to have pulses separated by up to 200 ms, R_H and L_H would have to be very large. In fact, it seems unlikely that the human body could have such a high impedance. In these cases, there are two possible causes for more widely separated multiple discharges. First, the dielectric absorption effect of the shoe soles has to be considered. One way of looking at this is to imagine that the shoe sole is an *RC* circuit, with a very large *R* (and thus a very long discharge time). This *RC* circuit is resupplying charge to the human body, that is C_H, C_A, and C_F.

The other possible cause of widely separated multiple discharges is related to the motion of the person toward the keyboard. As previously mentioned, the arc will quench when there is insufficient energy to maintain the air ionization path. The arc may then not restart until the person's finger has moved closer to the equipment, so the arc gap is shorter and less energy is required to initiate the arc.

If there is recharge of C_F and C_A (and possibly C_H), it will occur even if multiple discharges don't occur. This recharge (and potential multiple discharge) could affect electronic equipment. If the person recharges, the equipment (in the example, the keyboard) must also recharge. Because the recharge current may have a rise time on the order of microseconds or slower, most of the current would flow via the equipment ground path and not C_K. (C_K is typically only tens of picofarads or less.) For fast recharge, with microsecond rise times, the current flow in the ground could have an impact on the operation of the equipment, and could be one more source of relatively low frequency noise associated with ESD. Whichever explanation is used, multiple discharges are more likely when the person has a high initial charge level. Ironically, initial discharges in the sequence are likely to have lower peak currents and slower rise times (because corona is more pronounced at higher voltages) than subsequent discharges in the sequence. The phenomenon of multiple discharges may explain another effect noted by many experimenters. It has been noted that both low voltage and high voltage ESD will often cause more problems than medium voltage ESD. The key to explaining this may be that fast rise time, high peak signals cause the most problems. At low voltages, there is very little corona, and therefore rise times will be fast and peak currents high. At medium voltages, there is corona, which slows rise times and reduces peak currents. At high voltages there is also much corona, but multiple ESD becomes common. In each multiple ESD sequence there are one or more low voltage discharges which will have the fast rise times and high peak currents necessary to cause more severe problems.

2 ESD Effects in Electronic Equipment

In the previous chapter, a model for the electrostatic discharge (ESD) event was developed using the example of a person walking across a carpet and touching a computer keyboard. In order to discuss ESD effects, this simple example is continued. However, as was previously mentioned, the basic concepts apply not only for this specific example, but also for any ESD event. Although the example used only discusses a simple electronic system, most systems are much more complex. The reader must remember that solutions to ESD problems will require a system-wide approach, no matter how complex the system may be. During the ESD model development in Chapter One, the example computer keyboard was treated as a single block, with lumped capacitance, resistance, and inductance. In fact, this ''block'' is typically an enclosure, and within the enclosure are the switches and electronic circuits of the keyboard. If there is no other discharge current path or shielding available, the electrostatic field and discharge currents from the person will directly affect the electronic devices within the keyboard system.

The intensity of an electrostatic field is dependent on the charge levels, and on the distance between the items which differ in charge. It is very common for a person to have a voltage of 8 kV to 10 kV. Higher voltages such as 12 kV and 15 kV are rarer, but still possible. Many sources state that a human can hold charges of up to 30 kV or greater; however, this assumes that the smallest corona discharge radius of the body is 1 cm. In fact, the human body has radii smaller than 1 cm, so such a voltage level should not really be possible in normal conditions. A voltage of 20 kV is probably a more realistic maximum for the human body. (Clothing, hair, or shoes could hold higher charges than the body because they are typically much less conductive and, therefore, less affected by corona.)

In the example, as the person's finger nears the keyboard, the electrostatic field from the finger will cause polarization of the dielectrics and redistribution of charge within the keyboard. Charge redistribution in conductors will increase the polarization forces on the dielectrics and may even be so intense as to cause dielectric breakdown. This is especially possible within integrated circuits (ICs), which have thin dielectric barriers.

Although the electrostatic field alone can cause problems, an actual discharge causes even greater difficulty, because it adds the effect of direct charge injection. In this case, the energy in the field that previously existed between the person and the entire electronic system may now exist between the internal portions of components such as ICs. Chances for damage to the IC molecular structure thus increase drastically, due to dielectric breakdown caused by these very intense field strengths. Furthermore, currents following breakdown, or associated with charge redistribution, can melt or burn portions of the keyboard's electronic components.

One solution to charge injection problems is to place an insulating barrier between the person and the electronic components. This will prevent the actual discharge, as long as the dielectric of the barrier doesn't break down. Another solution is to place a metal barrier between the person and the electronic components. Of course, the metal barrier must be well isolated from the components, so it will not itself discharge to the components. A discharge will then inject charge into the metal barrier, not to the components. It can be seen for both barriers, however, that the electrostatic field problem would still exist. The only difference when the metal barrier is used is that the field after discharge is between the barrier and electronic system, instead of between the person and the system. (This may make the field more or less uniform, but doesn't make it go away.) To solve the electrostatic field problem, as well as prevent charge injection, it is necessary for the metal barrier either to completely surround the entire system (including cables), or to connect it to earth ground. An earth connection will bleed off the metal barrier charge and thus eliminate the electrostatic field, as well as prevent charge injection. (The opposing charge in the system will, of course, also bleed off to ground.) A metal barrier which *completely* surrounds the system will insure that no field can reach the system, even if the outside of the metal barrier is charged. (This is the case with a system in which all items of equipment have metal enclosures and all cables have shields.)

The approach of using metal barriers is often taken to protect equipment and is effective once installed and/or connected to ground. Unfortunately, it does nothing to prevent problems during manufacturing and handling. Current injection and field problems can still cause immediate or, even worse, latent defects. These ESD-caused latent defects result in field failures and can be a prime reason for "infant mortality" in equipment. Unfortunately, people can't feel fields or discharges less than a few thousand volts, so they aren't even aware of the damage they are doing. This problem will be discussed further in Chapter 8.

Returning to the example of the person and computer keyboard, assume the keyboard is protected during production and transport. Assume also that the example keyboard is beneath a metal plate, which is isolated from the electronics and connected to an earth ground. When the discharge occurs, currents will flow in this metal plate and ground line. These currents can be classified into two groups. First, charge redistribution currents will equalize the plate charge with the charge in the person's finger and arm. In this case

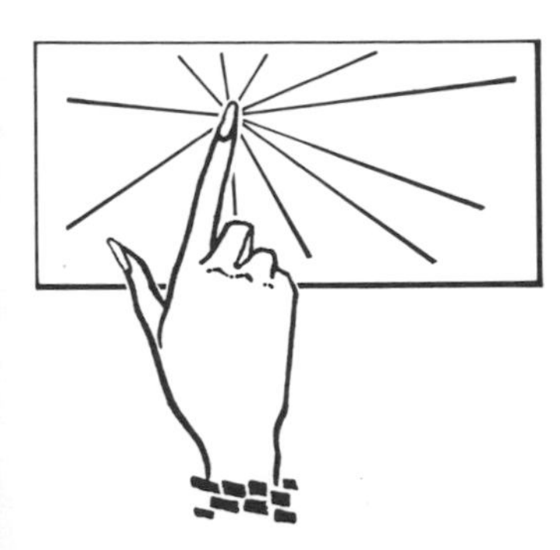

Fig. 10: *Radial current flow for ESD current.*

the plate is the source of charge. Second, currents in the ground path(s) will equalize the person's body charge with earth ground. The plate is not the primary source of charge for this second current, but merely the path by which currents flow. This can be physically related to the ESD model previously developed.

During the previous development of the ESD model, it was found that discharge currents associated with the capacitance of the hand, arm, and keyboard were responsible for the highest frequency components of the ESD pulse, and that these components were primarily responsible for generating intense fields. These high frequency currents are the redistribution currents within the plate. Also, in the model, discharge currents associated with the capacitance of the person's body to earth were responsible for the lower frequency components, and for carrying the majority of the energy of the discharge. These lower frequency components are the ground path currents. The physical path of the higher frequency plate redistribution currents will depend on the position of the person and of the plate, but will tend to be somewhat radial, as shown in Figure 10. This radial inflow (or outflow) of current occurs, of course, because the opposing charges stored in all portions of the arm and the plate are attracted to each other.

The physical path of the lower frequency body discharge current is not radial, but to ground, via the lowest impedance path possible. This is shown in Figure 11. Of course, real life is somewhat more complicated, but this explanation is substantially correct.

Now that the general paths and frequencies of the discharge currents are known, it is necessary to consider their effect on the electronic operation of the system. In this example, the low frequency currents are shunted to ground, and thus the keyboard and other portions of the system are protected from the destruction these high energy currents can cause. This protection from charge injection and destruction is, of course, the minimum requirement necessary for an acceptable design. However, the fields generated by these currents (especially the high frequency currents) will still have a significant effect.

Fig. 11: *Flow of ESD current to ground.*

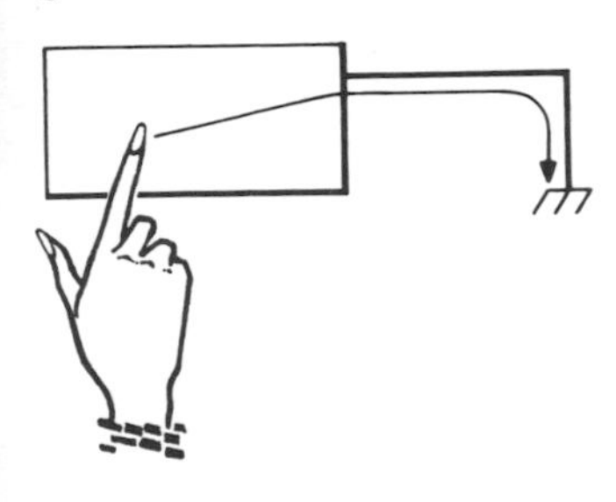

As the discharge currents flow within the system, they excite many antennas that exist in their path. The efficiency of radiation from these antennas is dependent primarily on the size of the antenna. The wavelengths of the frequencies resulting from an ESD pulse can range from centimeters to hundreds of meters. Since a one-quarter wavelength antenna is very efficient (even a 1/16 wavelength antenna can have significant radiation), antennas can easily have dimensions from 1.5 cm to 150 m. To see how these antennas are formed in an electronic system, the keyboard example will be used again.

Previously, it was stated that there was a plate over the top of the keyboard; however, a keyboard actually requires large openings for the key switches. (Very few real-world enclosures can be perfect shields because openings are almost always required. If nothing else, seams exist between parts.) In addition, there is usually a cable connecting to the plate, and it may have coils

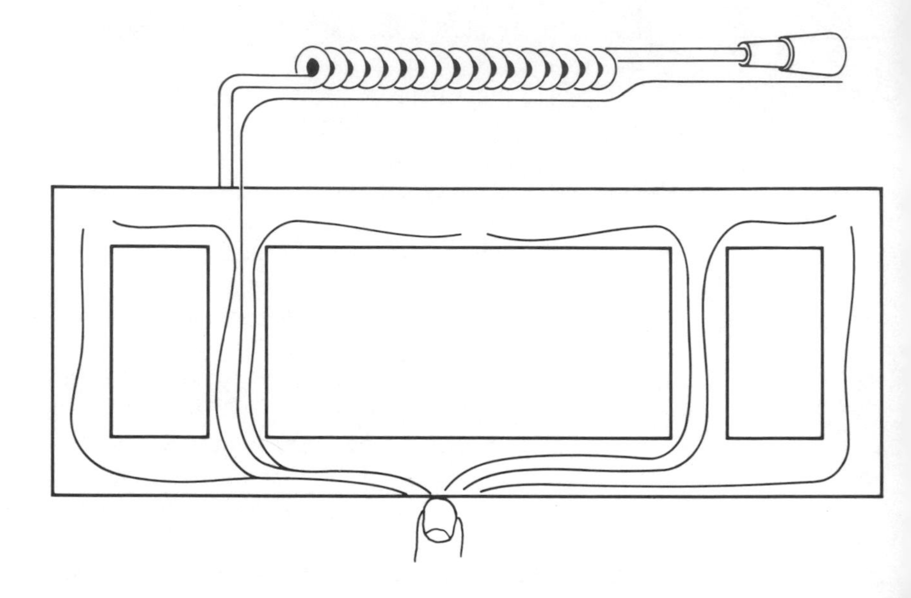

Fig. 12: *ESD current in the keyboard example.*

of some type, as shown in Figure 12. The resulting plate and cable connection to earth ground becomes a combination of loop, slot, and straight wire antennas. (More complicated systems, of course, have even more antennas.)

It's easiest to consider this combination one section at a time. Figure 13 is a diagram of the cable tangent (straight section). This will act as a straight antenna, and the fields would generally be oriented as shown. The field direction depends on whether the person is positively or negatively charged. Since this is not usually known, the field direction will be deleted in remaining diagrams.

Figure 14 is a diagram showing field orientation for a single loop of the coiled section of the cable. This is obviously a loop antenna. What's more, this loop antenna may be "tuned" to different frequencies by stretching the cable coils.

Depending on their dimensions, the openings cut in the plate may be considered as slot antennas, or the conductors between the slots may be

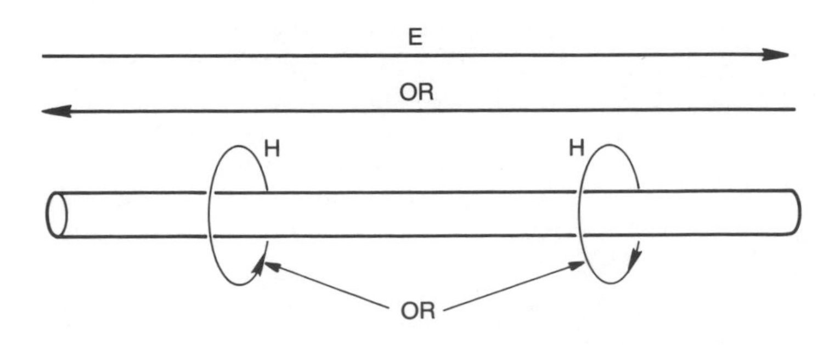

Fig. 13: *Fields for ESD current in cable tangent.*

considered independent antennas with fields like that of the cable tangent section (Figure 15).

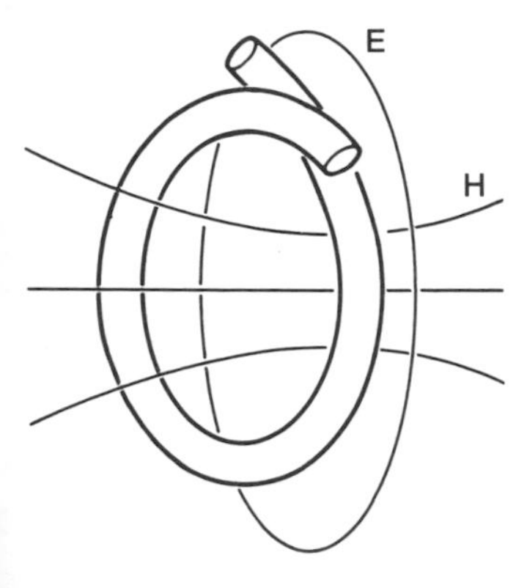

Fig. 14: *Fields for ESD current in cable coil.*

The above diagrams indicate that both electric (E) and magnetic (H) fields are generated. However, the relative energy of the E and H fields is not predictable in the region called the near field, which extends to about 1/6 wavelength from the antenna. At 1 GHz, 1/6 wavelength is 5 cm. This means most of the keyboard electronics are within the near field for most ESD frequencies, which is often the case in electronic systems. The relative strength of the fields is dependent on many factors, such as antenna gain, but, in general, a high impedance antenna radiates E-fields better, and a low impedance antenna radiates H-fields better. (Of course, this is expected because E is directly related to V, and H is directly related to I.) Both low and high impedance antenna structures will typically exist, so both types of fields will exist.

These fields, transmitted by the antennas of the plate and cable assembly, are received by antennas in the electronic circuitry of the system. As the lines of magnetic flux, which are caused by ESD currents, propagate outward from the transmitting antenna, they cut across the conductors in the circuits. This changing flux density generates an opposing current in the system conductors. Likewise, the transmitted E-field will induce a voltage differential in system circuits.

Just as high impedance antennas transmit E-fields better, so, also, do high impedance antennas receive E-fields better. Likewise, low impedance antennas receive H-fields better. This means, whether the system circuits have high or low impedance, they will be receptive to one or the other. H-fields will induce currents in low impedance loops, and E-fields will induce voltages on high impedance lines. Although each circuit is different, there are some impedance combinations that seem to be typical. Power and ground circuits are typically low impedance, input circuits are very often high impedance, and output circuits may be either high or low impedance. This means power and ground circuits will be susceptible to H-field induced currents, and steps must be taken to combat H-fields and/or the currents they generate. Inputs, on the other hand, typically will be more susceptible to E-fields and their induced voltages. In this case, the E-field must be fought, and/or voltage controls must be instituted. For output circuits, both types of

Fig. 15: *Fields for ESD current in a metal plate.*

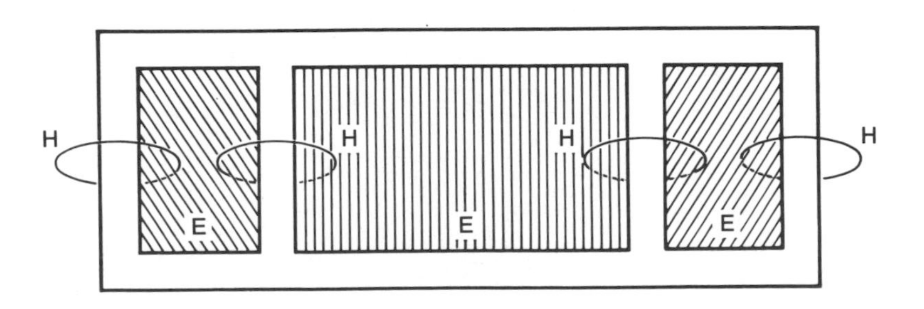

fields, and thus, both voltage and current effects may have to be considered. The symptoms of these effects may range from system lockup, to write and read errors, to incorrect characters appearing on a CRT display.

In analyzing the effects of ESD fields, it must also be remembered that fields can have indirect, as well as direct, effects on circuits within a system. This occurs when the field induces a current, or voltage, on a conductor, which conducts it to some point beyond the reach of the field itself. A good example is a field which induces a current on a cable shield. If the cable shield is not properly terminated, the induced current may pass inside an otherwise shielded enclosure. In this case, even though the original field could not penetrate the enclosure, the effect of the field will penetrate due to the induced current on the cable.

Also, just as noise fields can turn into noise voltages or currents, so can common mode noise be converted to differential noise. This is important to remember because truly common mode noise would usually have little adverse impact if only it could remain truly common throughout the system circuits. Unfortunately, what begins as common mode noise usually turns into differential noise at some point in the system circuitry because of unequal changes in amplitude, phase, or frequency content. For example, if every line of a cable isn't terminated in exactly the same way, the common mode noise seen at a cable connection will turn into differential noise in the receiver circuit.

Although ESD solutions must be applied to an entire system, it is difficult to discuss solutions without subdividing electronic systems into conceptually smaller groupings. In the chapters that follow, ESD solutions related to firmware, printed wiring boards, cables, and other system subdivisions are discussed. The reader must remember, however, that system problems cannot be solved by applying solutions only to portions of the system. A system will only be as immune as its weakest link. Therefore, solutions must be applied to all portions of a system.

3 Firmware/Software Design Guidelines

This is the first of several chapters discussing ways to reduce the effects of electrostatic discharge (ESD) on electronic systems. Many readers may be surprised that, in addition to the more well-known hardware methods of coping with ESD, firmware and software have an important role in the development of an ESD resistant system.

Although firmware cannot prevent ESD from damaging system components, firmware can drastically impact the nondestruction failures. Provided that the hardware and protocol allows it, nonrecoverable equipment failures (lockup) can often be eliminated, and even recoverable failures can be reduced by as much as 10 times, with properly written firmware.

As is the case with hardware ESD fixes, ESD resistant firmware has a cost. Typically, a program will be larger: this means that more programming time, and more memory to hold the program, are required. This cost must be compared with the cost of fixing the ESD problem using hardware alone. In many microprocessor applications, firmware fixes are more cost effective than hardware.

One of the key concepts in writing ESD immune firmware is mistrust. One should never assume that the state of a port, register, etc., has not changed. In fact, one should assume the opposite. If, for example, an index register is about to be used, one should ask what would happen if this index were wrong. If only a temporary and fairly mild problem would occur, such as light emitting diode (LED) flicker, then no special action may be required. If a more serious problem could occur, such as system lockup, then measures should be taken to prevent this.

Firmware (or software) ESD preventive measures fall into two general categories:

- Refreshing.
- Checking and Restoring.

In the following paragraphs, each category is discussed in concept and some specific examples are given. Since it is impossible to foresee all specific cases, it is important for the person generating the firmware to be constantly aware of the overall concepts in order to obtain firmware that is resistant to ESD problems. Although the following discussion deals primarily with firmware, many of the concepts are appropriate for software as well.

Refreshing

When refreshing, the programmer is not concerned about the past history, but only with insuring things are right for the next step. For example, prior to reading data with an 8049 microprocessor port, ones should always be written to the port lines. This should be done even if the port has not been modified since the last input. Do not assume the port still has ones written to it. Other items for which refreshing should be considered are:

A. Re-enabling interrupts at regular intervals (this would include executing RETR in an 8049 and RETI in an 8051).
B. Refreshing stop bit levels when ports are used for serial data outputs.
C. Refreshing latch and port output status.
D. Reading control and selection inputs (e.g., from jumpers or switches) at regular intervals to insure the system is operating in the proper manner.
E. For 8049 and 8051 processors, there should be at least one register bank select instruction contained in every program loop.

One other consideration, when refreshing, is the order in which items are refreshed. Sometimes one item must be refreshed before another. For example, on a synchronous I/O, the rest state of the data line should usually be refreshed before the rest state of the clock line. Otherwise the clock refresh may actually clock out a data bit. As previously mentioned, you should always be thinking about the consequences of each instruction under error conditions.

Checking and Restoring

Sometimes it isn't sufficient to merely refresh. In some cases refreshing could even mask serious problems. In these cases the register, port, etc., should be checked to determine its state. If the state is improper, the program should attempt to restore the proper state.

Restoring (or re-initializing) must be done carefully. Although the state of the system may be suspect, it is usually not a good idea to totally clear it and start from scratch. This would result in the loss of all past history. The guideline should be to place the system into the most likely state, *and* in the state which will cause the least severe problems. If some critical item has no "most likely state," then its status should be stored redundantly. This will allow a vote to be taken to determine the present status. Usually a "2 out of 3" vote is sufficient. If no two of the redundant items compare, then a default condition is chosen.

Checking functions can generally be grouped into three categories. Specifically, the following should be checked, and if they are wrong, a restoration should be done:

A. Check that program flow is correct.

a. Subroutine stack pointers may be checked prior to execution of a return, and periodically in the main program. This is done to verify that the program subroutine nesting is within the expected range.
b. Instead of, or in addition to, checking the stack pointer value, use "tokens" to help detect program flow problems. When entering a subroutine, store a token, then, when exiting the subroutine, check for this token.
c. Place "trap" codes in forbidden areas, such as code tables or unused interrupt vectors. If a program tries to execute these codes, it will be caught. (A good example is to place return instruction op codes in unused table locations.)

Two other routines should be included in a program to allow it to check its flow:

d. A timer interrupt routine that is never stopped or disabled by the program should be used to verify that the main program is operational.
e. Periodically, the main program should check the above timer to verify it is still in operation. (If there is not a built-in microprocessor timer, then an external hardware "watchdog" circuit can be used. If the processor is not periodically set, this external circuit should reset the processor.)

B. Check that stored values and information are correct.

a. Periodically vote on redundantly stored items, and if there is no agreement, re-initialize. In specific, status flags (especially enable/disable flags) should usually be stored redundantly. Error correcting codes may be used in place of, or in addition to, redundancy.
b. Index and other important registers should be checked for proper value or range prior to their use (especially if writing to RAM).
c. If there is too much critical information to use simple redundancy, and if it isn't possible to check all information for proper values, then use checksums, and/or cyclical redundancy checks (CRC) to check blocks of data.

C. Check that inputs and outputs are correct.

a. Check inputs for proper parity, framing and/or checksums.
b. Verify that inputs are reasonable. Some received codes may be clearly in error.
c. Sample all input levels at least twice to perform "software filtering" of noise.
d. Outputs can be checked by having the receiver echo them.
e. A receiving device should acknowledge all valid inputs, and the sending device should retransmit if there is no acknowledgment.

If any of the above checks are failed, there should be a restoration. It should be apparent that a restoration cannot usually be the same as the initialization routines utilized during power on (or hardware) reset. For example, RAM shouldn't be cleared. In fact, the processor should be prevented from executing the hardware initialization routines in the event of a false program counter reset caused by ESD. This can usually be done by checking a flag register prior to execution of hardware reset routines. If the flag is set, a full reset would be prevented. This flag would be set when the main program is executed, and would normally only be cleared by actual hardware reset of the processor.

In specific, restoration should often do the following:

1. Reset subroutine stack pointer.
2. Reset FIFO pointers.
3. Reset counters.
4. Prevent transmission of suspect codes.
5. Disable interrupts until restoration is complete, then re-enable and restart timer.
6. Reset interrupt pending flags.
7. Refresh outputs.
8. If the host system will accept it, it is a good idea to send the host system a code to tell it that there has been a restoration. The host could then take steps to insure that all portions of the system are in agreement.
9. And, of course, the restoration routine must clear the specific problem that caused the restoration.

Most of the preceding discussion assumed that the system was executing the program (although not properly). If the program includes tables, it is possible for the processor to try to execute these table values as instructions. It is theoretically possible for the table values to cause the watchdog timer to be disabled and then cause the processor to execute an endless loop. If a specific design has this problem, there is a solution. It is possible to find the offending lockup loop with a logic analyzer connected to the address bus. One of the table values within the loop should then be replaced by a return instruction. This will usually get the processor out of the loop. The program can then be changed so the one replaced table value is handled in a special manner. (If possible, an unused table location should be used to hold this return op-code. In fact, as previously mentioned, it is probably a good idea to fill all unused table locations with return op-codes.) This logic analyzer technique can, of course, find lockup loops in other areas of the program as well. However, if the design utilizes a single chip microprocessor with a masked program, then no address bus will be available. In this case, the program may be "ESD hardened" on the emulation system after the debugging process. Once the program is debugged and is functioning properly, begin changing registers at random points in order to simulate ESD effects.

The program counter is especially critical and should be set at random values. Also, the subroutine stack pointer should be set to various values. By making these changes in emulation and noting the results, many potential problems can be found. This may not always be easy, but systems with excessive problems are not acceptable. By applying these concepts, ESD problems related to firmware should be made relatively rare.

4 Printed Wiring Board Design Guidelines

This and several following chapters discuss hardware solutions for system problems caused by electrostatic discharge (ESD). To begin discussing system hardware solutions, it is useful to divide the ESD effects on systems into three categories:

- Effects due to electrostatic fields that exist prior to the discharge.
- Effects due to the charge injection from the discharge.
- Effects due to fields generated by discharge currents.

Although the design of printed wiring boards (PWBs, also often called printed circuit boards, PCBs) can impact all three types of effects, PWB design mainly impacts the third effect category. In the following discussion, PWB design guidelines are given to combat the third category problems.

In general, reduction of field coupling from a source to a receiver can be done in one of the following ways (these general methods will also be referred to in other chapters in which fields are discussed):

1. Use filters to attenuate the signal at the source.
2. Use filters to attenuate the signal at the receptor.
3. Reduce the coupling by increasing separation.
4. Reduce coupling by decreasing the antenna efficiency for source and/or receiver.
5. Reduce coupling by orienting the receiver antenna polarity 90 degrees off from the transmitter antenna.
6. Place a shield between the transmitter and receiver antenna.
7. Reduce the *E*-field coupling by decreasing the impedance of the transmit or receive antenna.
8. Reduce the *H*-field coupling by increasing the impedance of either antenna.
9. Couple signals so they remain common mode by using a uniform, low impedance reference plane (as may be provided by a multilayer PWB).

In specific designs, where either *E* or *H* fields predominate, methods 7 and 8 may work. In general, however, ESD generates both *E* and *H* fields. This means method 7 will result in an improvement in *E*-field immunity, but reduction in an *H*-field immunity. Method 8 has the opposite result. Therefore, methods 7 and 8 are not considered general solutions. Methods 1

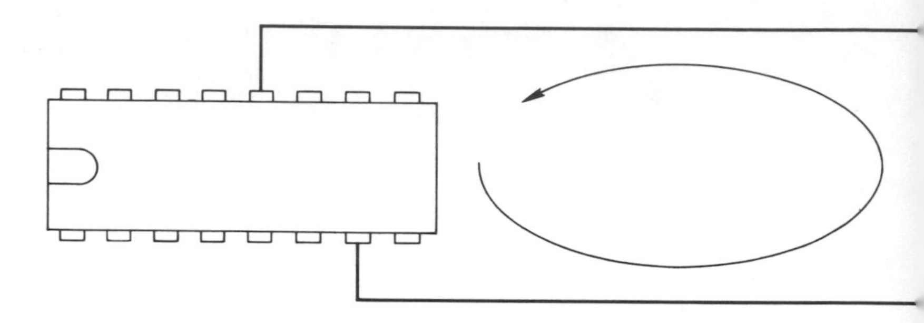

Fig. 16: *Simple PWB circuit loop.*

through 6, and 9 will work, no matter whether the field is E or H, but PWB design solutions rely heavily on methods 3 through 6, and 9.

During the following discussion, six practical rules for achieving solutions via methods 3 through 6, and 9 are detailed along with an explanation of the reason for each rule.

I. Minimize Loop Areas

Any circuit loop which encloses a changing magnetic flux will have current induced in the loop. The strength of the current is proportional to the amount of flux enclosed. A smaller loop will enclose less flux, and thus have less current induced. Thus loop area must be minimized.

The difficulty encountered in applying this rule is to recognize loops. Anyone will recognize the loop shown in Figure 16, but it is more difficult to recognize the loop shown in Figure 17.

Rather than attempting to identify all possible loops, it is recommended that the following specific steps be taken to prevent large loops:

A. Power and ground traces should be kept close together to reduce power to ground loop area. The examples in Figure 18 show different methods of routing power and ground to ICs.

B. Multiple power and ground traces should be connected in a grid. Figures 19 and 20 illustrate what this means: In this typical PWB design example, one side of the PWB has the vertical lines and the other side has the horizontal lines. (Only the ground lines are shown in this case.)

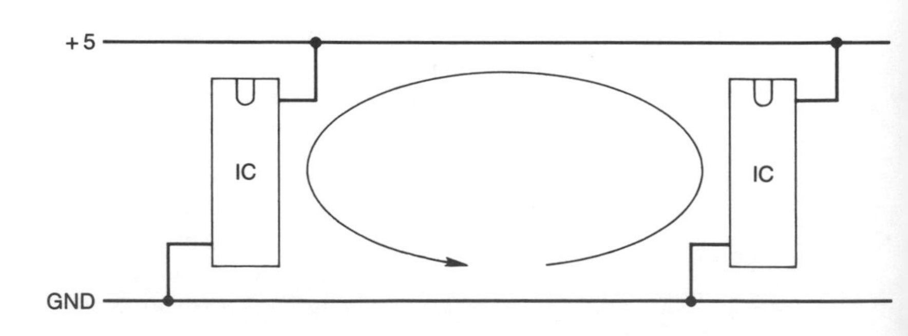

Fig. 17: *PWB loop formed by power and ground.*

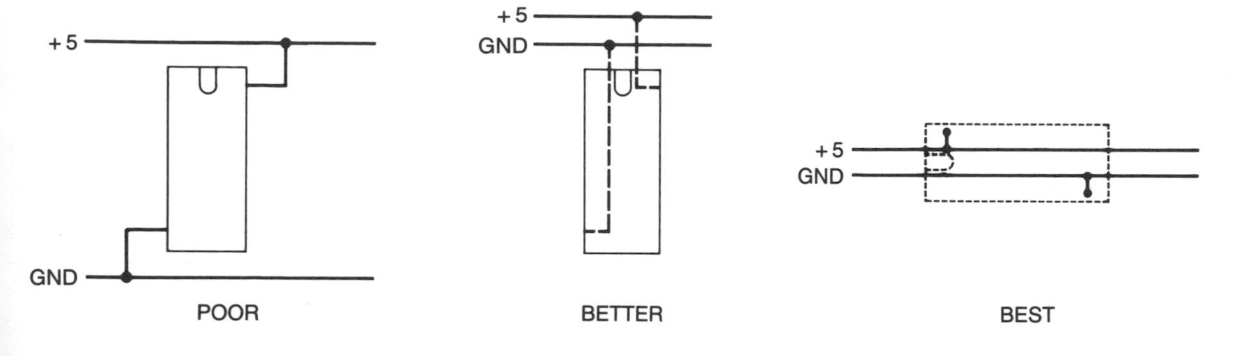

Fig. 18: *Reduction of power and ground loops.*

As Figure 19 illustrates, this typical ground connection results in large loops. These loops can be reduced by adding more feedthroughs to connect traces on the two layers, as shown in Figure 20. The loops formed by the grid have much smaller areas. This results in lower induced currents, and fewer problems. The grid concept can be applied in three dimensions also. Those PWBs that are plugged into backplane (or mother) PWBs should have multiple ground and power connections that are equally spaced along the length of their connectors. This will result in smaller loop areas for the system as a whole.

Fig. 19: *Typical PWB ground.*

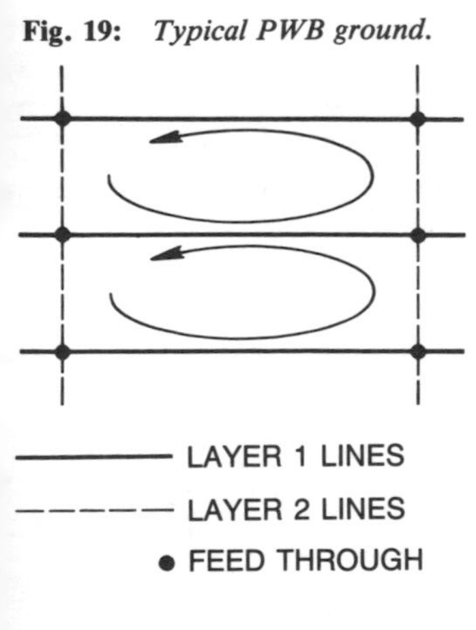

Both A and B above will reduce the efficiency of loop antennas associated with power and ground loops. Items C and D, which follow, will reduce the efficiency of loop antennas involving signal lines.

C. Conductors connected in parallel must be kept close together, or, better yet, only one large conductor should be used. This also means a ground plane shouldn't have large holes. Such holes are parallel conductors and act as loop antennas.

D. Signal lines should be kept close to ground lines. A logical step would be to run a ground line next to every signal line. Unfortunately, this may result in many parallel ground lines. To circumvent this problem, use ground plane or grid, as previously mentioned, instead of individual ground traces. An example is shown in Figure 22. The versions in Figure 22 assume that, for some reason, the signal line could not be moved.

A ground plane could also be placed on the PWB opposite the signal line, as shown in Figure 23. In fact, it is a good idea to fill all unused PWB areas with ground plane.

E. Long sections of power or signal lines between especially sensitive components should be transposed, at regular intervals, with ground. Transposing the lines means simply changing one of the lines from top to bottom, or left to right and doing the opposite with the other line. Figure 24 illustrates that this has the same effect as reducing the loop area: after transposing lines, only small loop areas exist.

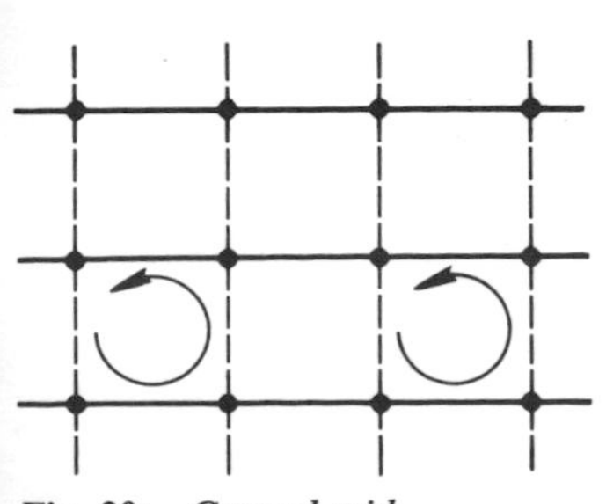

Fig. 20: *Ground grid.*

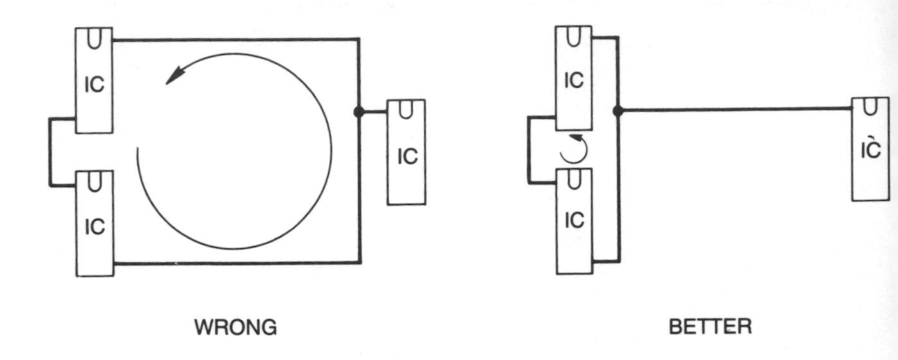

Fig. 21: *Reduce parallel paths.*

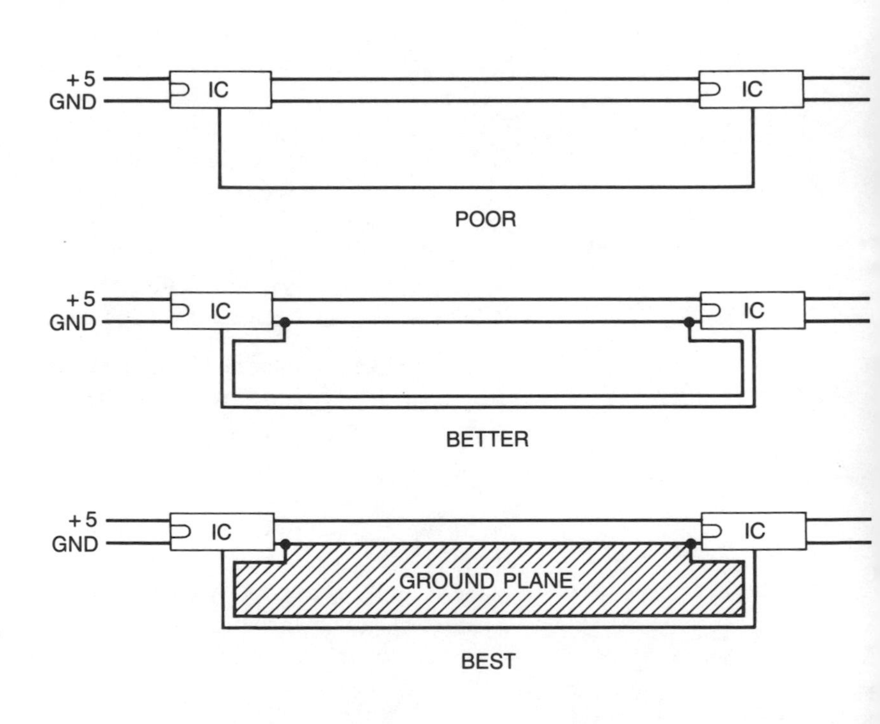

Fig. 22: *Keep signal traces close to ground.*

Fig. 23: *Signal traces over ground traces or plane.*

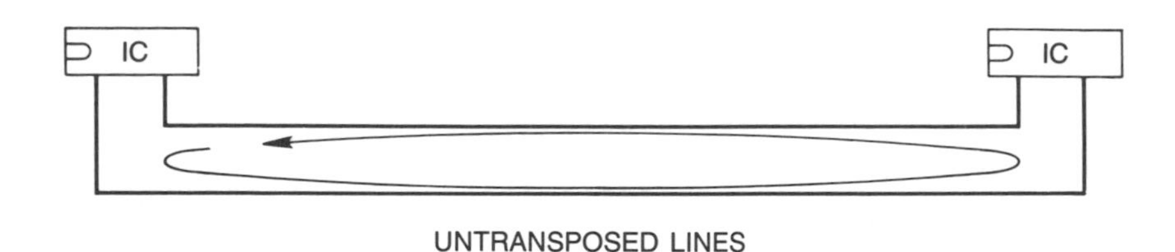

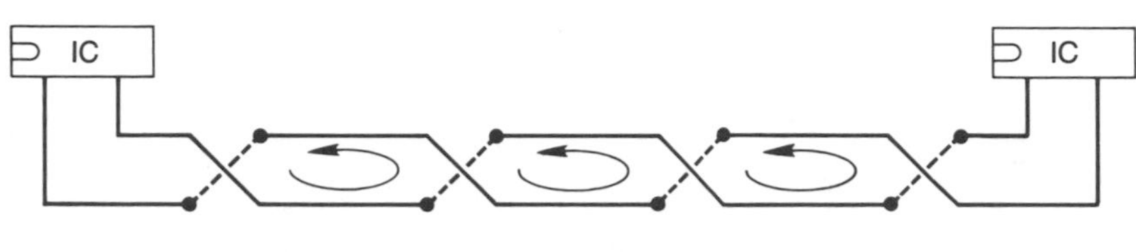

Fig. 24: *(top) Untransposed lines; (bottom) transposed lines.*

F. Install *high frequency* bypass capacitors between power and ground. Since the bypass capacitors have low impedance at the lower ESD frequencies, they reduce the loop area of power and ground at these frequencies. At high ESD frequencies, however, even high frequency capacitors are of limited use because of stray inductance.

Of course, the closer the power and ground lines are to each other, the less noticeable the effect of the filter capacitors, because the loop area is already small. Figures 25 and 26 illustrate this effect. Even with the bypass capacitors beside each component, the circuit in Figure 25 still has a large loop area.

Fig. 25: *Larger loop associated with bypass capacitor.*

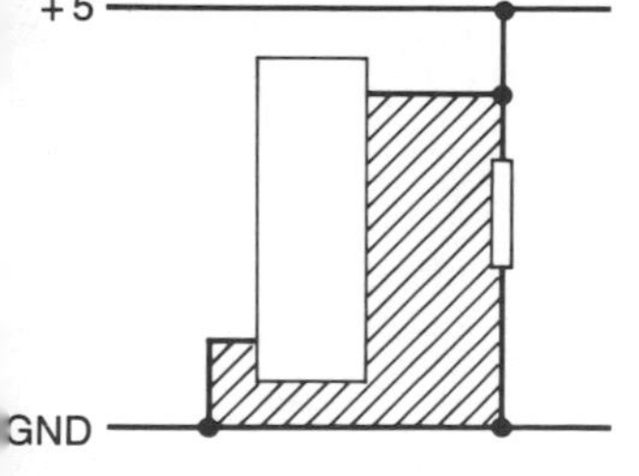

The circuit in Figure 26 reduces the loop area greatly by putting power and ground close together. However, even when power and ground are side-by-side, long sections of lines can result in a fairly large loop area.

This discussion of long lines leads to the next general PWB design rule.

II. Keep Line Length Short

In order to be an efficient antenna, a line must be a significant fraction of a wavelength. This means a longer line will efficiently receive more of the frequency components generated by an ESD pulse; a shorter line will be an efficient antenna for fewer frequency components. Therefore, a short line will receive less energy from ESD-related fields to feed into the system circuits.

Fig. 26: *Smaller loop associated with bypass capacitor.*

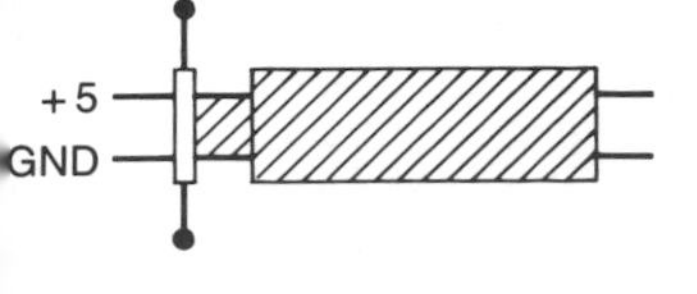

Keeping lines as short as possible is a more commonsense process than keeping loops small. Unlike a loop, which may not be immediately apparent, line lengths are clearly visible. Some specific design steps follow:

A. Keep all components close together. The PWB designer shouldn't spread components to use extra PWB area.

B. Within a group of components, those components with the most interconnects between them should be the closest to each other. This would

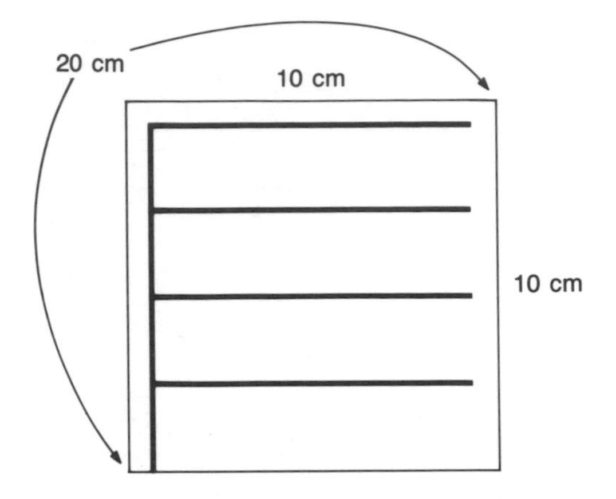

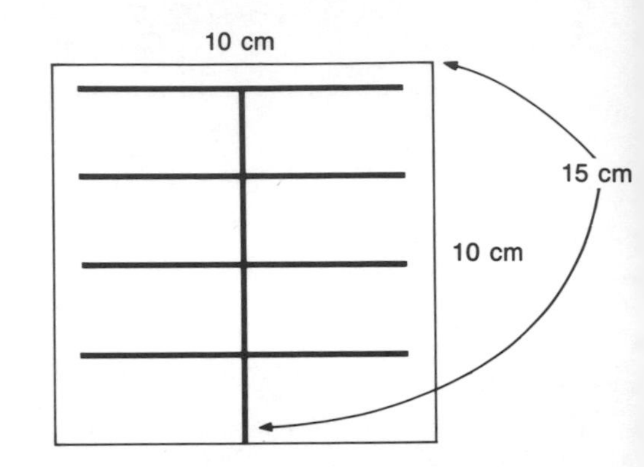

Fig. 27: *(left) Corner fed signal; (right) center fed signal.*

mean, for example, that I/O components would be the closest components to the I/O connector, etc.

C. If possible, feed power or signals from the center of the PWB edge rather than from the corner. As Figure 27 shows, a center fed signal results in the shortest lines to the most components. The greatest line reduction will result when square PWBs are used. Long, narrow PWBs will show little, if any, line reduction with center fed signal or power lines.

The previously indicated PWB design guidelines are primarily concerned with effects due to fields generated by ESD currents. It is interesting to note that all the previous steps, which were intended to reduce antenna efficiency, will also help prevent common mode noise from becoming more troublesome differential noise, as was mentioned in item 9 of the general methods at the beginning of this chapter. This is the case because all the previous steps tend to reduce the differences in various PWB path impedances. For example, rule I-step D is especially helpful because it keeps the signal and associated ground return impedances nearly equal. Therefore, common mode noise that is introduced into these two paths will remain nearly equal in amplitude, and little signal-to-ground differential noise will develop. In addition, PWB design can also provide some help in reducing problems due to electrostatic fields, and to charge injection. The following design rules relate to these problems. You will note that several of these rules duplicate the previously mentioned rules.

III. Fill PWB Areas with as Much Ground Plane as Possible (Multilayer PWBs are Very Helpful)

It was previously mentioned that ground plane helps reduce loop area which thus reduces receive antenna efficiency. Ground planes also help reduce electrostatic field problems by acting as the primary source of charge to oppose the charge on the ESD source. This means less charge must reside in sensitive components and signal lines. The PWB ground plane also acts as a shield for circuit lines on the opposite side of the PWB. (Of course, the larger the holes in the ground plane, the less effective the shielding will be.) In addition, if a discharge should take place, and the ground plane area on the surface of the PWB is large, there is a better possibility that the charge will be injected into ground, and not into signal lines. This will reduce the chance of component failure because the charge may bleed off before causing damage. (However, even a discharge into ground can cause damage, and should be avoided.)

IV. Power and Ground Lines Must Be Well Coupled by Capacitance

Coupling between power and ground is accomplished in two ways, as previously mentioned.

A. Keep power and ground traces close together or use multilayer PWBs. This creates more stray capacitance between power and ground.

B. Use high frequency bypass capacitors to couple power and ground. (Combinations of capacitors may be used to couple both lower and higher ESD frequencies.) Coupling between power and ground will help reduce charge injection problems. The voltage differential, caused by the difference in charge levels on two objects, is dependent on the capacitance between the two objects ($V = Q/C$). If a charge of X is injected into the power supply line, it causes a voltage of Y between power and ground. If the capacitance between power and ground is doubled, a charge of X will only cause a voltage of $Y/2$. Of course, this smaller voltage will result in a lower possibility of damage.

V. Isolate Electronic Components from the Source of the ESD Charge

During the discussion of ESD effects, it was indicated that charge injection into the electronics can be prevented by isolation. With regard to PWB design this typically means physically separating the electronics from potential charge sources and from connector ports or signal lines where induced currents tend to concentrate. Two steps to achieve such separation are:

A. Keep electronic components and PWB lines away from PWB areas which are subject to ESD. (For example, areas the operator can touch directly.)

B. Keep electronic components and PWB lines away from any *metallic* objects that are subject to ESD. (This includes screws, brackets, connector shells, etc.)

This last requirement relates to the next design rule.

VI. Chassis Ground Lines on the PWB Must Be Low Impedance and Well Isolated

Although PWB soldermask can help isolate PWB traces, soldermask can have pin holes which allow arcing.

A. The best method to isolate chassis lines is to place them away from the electronics. In addition, if the chassis line presents a low impedance to the ESD current, the charge will bleed off without arcing. Of course, such a rapid bleed-off of charge will create stronger fields, but that is preferable to direct charge injection caused by arcing into the electronics.

B. A chassis ground trace must have a length-to-width ratio of 4, or 5, to one. A trace wider than this will reduce impedance (inductance) only slightly, but a narrower trace will increase impedance greatly. This length-to-width ratio naturally means that chassis ground traces must be short. Otherwise, their width will quickly become excessive, as length increases.

Overall Priority of Design Rules

This completes the discussion of PWB design rules as they relate to ESD. There are, of course, times when all the rules cannot be completely satisfied. When this is the case, a conscious decision must be made about what must be sacrificed. The three categories of potential ESD effects identified at the start of this chapter are useful to set priorities for prevention. In general, the following order of priority should be followed:

1. Prevent charge injection into system circuitry because this results in destruction.
2. Prevent problems due to fields generated by discharge currents.
3. Prevent electrostatic field problems.

Fortunately, most of the rules are compatible and all problems can be well addressed in a typical PWB design.

Summary of PWB Design Guidelines

The following twelve guidelines are listed, in order of priority, for ESD problem prevention:

1. Noninsulated chassis ground on the PWB must be separated from other traces by at least 2.2 mm. This applies to anything connected to chassis ground, as well as traces.
2. Chassis ground traces must have a length-to-width ratio of no more than 5:1.
3. Keep noninsulated electronics at least 2 cm away from PWB areas that an operator can touch, or nonchassis grounded metallic objects that the operator can touch. (Enclosure design must also be considered to meet this requirement.)
4. Power and ground traces must be kept either side-by-side on the same PWB layer, or one over the other on opposite PWB layers.
5. Ground plane and ground traces must be connected to form a grid. There must be vertical ground lines connected to horizontal ground

lines at least every 6 cm in either direction. Typically, on a double sided PWB, this means layer two may have vertical ground lines, layer one may have horizontal ground lines, and there must be a feedthrough at least every 6 cm to connect the two. (Of course, connections at intervals of less than 6 cm are even better, and ground planes are better than ground grids.)

6. All signal lines must be within 13 mm of a ground plane or line. The ground can either be on the same layer or the next layer above or below the signal line. If the signal line is 30 cm long or more, it must be directly beside a ground trace, or over a ground trace or plane on another PWB layer.
7. Bypass capacitors between power and ground must be no more than 8 cm apart. (This may result in more than one bypass capacitor per IC.)
8. Components with the most interconnects between them must be side-by-side, or end-to-end.
9. All components must be as close as possible to the I/O connector. (Remember, guideline 3 takes precedence over this.)
10. Fill all unused portions of the PWB with ground plane. (Remember to create a grid by connections at least every 6 cm.)
11. If possible, feed power or signals from the center of the PWB edge, not from the corner.
12. Long sections (30 cm or more) of especially sensitive signal lines should be transposed with their ground line.

Note: These design guidelines must be applied to the entire structure of PWBs within a system (e.g., backplane and attached PWBs). When applying guideline 2, for example, the length of the chassis ground trace includes its length on both the mother and daughter PWB.

5 Cable Design Guidelines

As discussed in Chapter Four, electrostatic discharge (ESD) causes electrostatic field effects, charge injection effects, and effects due to fields generated by ESD currents. Proper system cable and enclosure design can help combat all three types of effects. In this chapter, system cable solutions for ESD problems are discussed.

When discussing system cable solutions, it is convenient to reorganize the three ESD effects, mentioned above, into two categories:

- Effects due to radiated noise.
- Effects due to conducted noise.

Radiated noise includes the effects of both electrostatic fields and the E and H fields generated by discharge currents. Conducted noise includes direct charge injection and currents induced by E and H fields. In true life, of course, these effects are not isolated, but the discussion is simplified if only one type of effect is considered at a time.

First, cable guidelines to prevent, or reduce, conduction problems are discussed. At 20 kV an arc can jump 2 cm through air. Therefore, in order to prevent charge injection into the electronic circuit, a designer has three choices:

1. Design the equipment so an operator cannot physically come within 2 cm of the electronics or touch ungrounded metallic objects that are within 2 cm of the electronics.
2. Insulate all electronics within a better insulator than air. (Such insulation would have to have seams that were gas tight or were designed to be longer than 2 cm.)
3. Provide a target other than the electronics for the charge injection.

Solutions 1 and 2 above are discussed primarily in Chapter Six, since they are more related to the enclosure design. Solution 3, however, is heavily dependent on the design of the system cabling. As was discussed in Chapter Two, the ESD target must provide a path to ground to eliminate electrostatic fields.

This ground path must be, and remain, low impedance. Otherwise, there may be arcs to alternate lower impedance ground paths via the electronic circuit. In order to have low path impedance, the first requirement is that resistance of the system cables be low. For most systems, cable wire resistance is reasonably low, except at very high frequencies, where the "skin effect" will increase the effective wire resistance. The solution to the skin effect problem is to increase the surface area of the cable conductors. (This will be further discussed when cable shields are discussed below.)

Another way in which cable resistance may be increased is contact corrosion at connection points. Corrosion at contact points can create enough resistance to have as much impact as skin effect. In order to prevent corrosion and maintain a low resistance cable system, the following guidelines should be considered.

I. Materials in Contact with Each Other Should Be Close in the Electrochemical Series

Galvanic corrosion occurs when metals with different electromotive force (EMF) levels are in contact in the presence of moisture. The level of corrosion is dependent on the difference in EMF levels of the materials. The greater the difference, the more the corrosion. Table I lists a partial electrochemical series. It should be noted that the EMF listed in each case can vary somewhat. For example, beryllium is sometimes placed below aluminum.

The difference in EMF that can be tolerated depends on the environment. In severe marine environments there should be no more than 0.25 V difference between materials. Normally, except for connections to aluminum, it should be no problem to keep the difference under 0.75 V. For example, a zinc plated steel chassis could easily be connected to a copper ground wire via a brass lug.

II. The Anodic Material (More Positive) Should Be Larger than the Cathodic Material

It is the cathodic material which provides the electrons for corrosive action at the anode. A large cathode and small anode will thus result in worse corrosion than a large anode and a small cathode. There is a corollary to this requirement for a large anode-to-cathode ratio based on exposed area. Finishes, such as paint, on the anode or cathode reduce their exposed area. The *unfinished* area of the anode must be larger than the *unfinished* area of the cathode.

III. The Contact Shouldn't Have a Constant Current Flow

In addition to the previously discussed galvanic corrosion, there is electrolytic corrosion. Electrolytic corrosion occurs whenever current flows from one metal to another via an electrolytic solution. (A moist power connection fulfills the requirements for electrolytic corrosion.) Electrolytic corrosion occurs even between metals with no inherent EMF difference. However, electrolytic corrosion doesn't occur if there is no current flow. A

TABLE I: *Electrochemical Series*

Metal	EMF (Volts)	Metal	EMF (Volts)
Magnesium	+2.37	Lead	+0.13
Magnesium Alloys		Brass	
Beryllium	+1.85	Copper	−0.34
Aluminum	+1.66	Bronze	
Zinc	+0.76	Copper-Nickel Alloys	
Chromium	+0.74	Monel	
Iron or Steel	+0.44	Stainless Steel	
Cast Iron		Silver Solder	
Cadmium	+0.40	Silver	−0.80
Nickel	+0.25	Graphite	
Tin	+0.14	Platinum	−1.20
Lead-Tin Solders		Gold	−1.50

Source: EMC Handbook, vol. 3, Don White Consultants.

chassis ground or shield connection is an obvious case of a contact which should have no constant current flowing through it, so this guideline should be easy to follow.

IV. Use Cathodic Materials, If Possible

Cathodic materials, such as gold, are more stable than anodic materials. The atmosphere itself can easily cause oxidation of highly anodic materials such as aluminum. Gold, on the other hand, is well known for its resistance to oxidation. Clearly, there are other tradeoffs involved in the selection of materials, but the designer should be sure to consider oxidation when making a materials decision.

If corrosion and skin effect resistance are controlled, the primary impediment to the flow of ESD currents will be the transmission line effects of the cable at the high ESD frequencies. Ideally, exact impedance matching could be used to insure that system cables have the minimum impact on the flow of ESD energy to ground. In reality this isn't possible. A system chassis ground includes not only the interconnect cables of the system, but also the safety ground in the AC wiring of the building, etc. The impedance of the total ground path is subject to impedance mismatches at various connection points. This is why it is not really possible to make use of impedance matching.

In spite of this limitation, however, a low resistance ground path can still act to bleed off a significant amount of the ESD charge, and thus prevent it from arcing into the electronics. This is so because even a completely open circuit line termination will look like a short if the line input is an odd number of quarter wavelengths away. Likewise, a short circuit line termination looks like a short if the line input is one or more half wavelengths away. Therefore, regardless of the type of termination mismatch that occurs in the ground path, many frequency components will see a relatively low impedance, because they will be the proper number of fractional wavelengths from the termination.

Other frequency components, however, will see a relatively high impedance. As previously mentioned in Chapter 4, this means the system chassis ground path must be isolated from the system electronics. Otherwise, there will be arcing from the chassis ground path to other lines within the system. One obvious means of isolation is the cable insulation; however, at connection points there typically will be air gap paths from chassis ground to other circuit lines. (Connectors usually don't have gas-tight seams.)

V. Air Separations from Chassis Wires or Connector Pins to Other Lines Should Be At Least 2.2 mm to Prevent Arcing

Unfortunately, for most inexpensive connectors the separation between pins is less than 2.2 mm. The 2.2 mm rule-of-thumb is conservative, therefore, a less expensive connector may be acceptable. For most ''D'' connectors and DIN connectors there is no problem because the shell is about 2.2 mm from the pins, so chassis ground can easily be connected to the shell. An alternative is for the chassis ground line to have its own connector, completely separate from other connectors. In this case, the 2.2 mm separation is easy to achieve.

Once steps have been taken to reduce problems due to conduction, problems associated with radiation must be addressed. The first radiation problem encountered is the radiation from chassis ground itself. The impedance mismatches that exist in the system chassis ground path create standing waves. The standing waves result in both E and H fields distributed along the system cables, and some means must be found to reduce the effect of these fields on other signal lines within the system.

In Chapter 4, nine methods of reducing antenna coupling were presented. Of those nine methods, only method six is of much use in isolating cable lines. Method six is the insertion of a shield between the transmitter and receiver. This leads to the next cable design guideline.

VI. Use Shielded Cables and Connect Chassis Ground to the Shield

Because of skin effect, the ESD current will tend to flow on the outside surface of the shield. (This is what makes a wire appear to have a larger resistance at high frequencies.) A shield will have more surface area than a normal wire, and will thus reduce the ''skin effect resistance.'' Also, assuming the shield is sufficiently thick, the ESD current will flow along the outer layer of the shield, and the inner layer will still be a shield. The inner layer of the shield will, therefore, act to reduce fields generated by ESD current before the fields reach the other lines inside the cable.

VII. Cable Shields Should Be At Least 0.025 mm Thick (Over the Range of 1 MHz to 5 GHz, Copper or Aluminum Doesn't Have to Be Very Thick to Provide Shielding)

Once the effort has been made to add a shield to a cable, don't waste it. Many designers are concerned about creating the infamous ''ground loop.'' They therefore insist on connecting cable shields at only one end. Unfortunately, while this may help reduce low frequency noise problems (such as 60 Hz), this will obviously make the shield ineffective as an ESD discharge path. This leads us to the next cable requirement.

VIII. Cable Shields Must Have a High Frequency Connection to the Chassis on Both Ends of the Cable

A. If no ground loop will result, or a ground loop won't be a problem (see note below), the best connection is a metallic connection at both ends. An example, in which no ground loop will typically result, is a cable going from a terminal to a keyboard.

B. If a ground loop will result, and will be a problem (see note below), then connect the shield to the chassis at one end by a metallic connection, and connect the shield to the chassis at the other end via a high frequency capacitor. An example of this case is the interconnect cable between a computer and printer. The key point, of course, is that both the printer and computer are connected to the safety (''green wire'') ground via an AC outlet. This could form the ground loop from one AC outlet to the next.

Note: Ground loops are not necessarily a problem unless they result in voltage differentials on the ''ground'' of items of equipment which are connected together. In general, the better the cable shield terminations, the less the probability that ground loops will become a problem.

IX. The Cable Shield Should Connect to the Chassis at the Cable Entry Point, and the Unshielded Portion of the Cable Must Be Kept to a Minimum

Remember that the exterior of the shield is itself radiating noise due to the ESD current. A shielded cable that runs close to a PWB, or inside a shielded enclosure, is a transmitting antenna to everything on the PWB or inside the enclosure. A cable shield only shields the wires inside it from ESD noise. It actually radiates ESD noise to everything outside it. Likewise, a long ''pigtail'' which serves to connect the cable shield to the chassis is also a transmitting antenna. Don't loop it around sensitive inputs. Of course, another problem with long pigtails is that they increase the impedance of the shield connection and thus reduce shield performance.

But what of those system designs which have no real chassis to connect to? For example, what can be done about a plastic enclosed keyboard, with no metal chassis, that is connected to the host terminal by a six foot cable?

X. If There Is No Chassis at One End of the Cable, Design In an Option to Connect the Cable Shield to Logic Ground via a High Frequency Capacitor

Designing a unit with no chassis connection point is not a good idea (as is discussed in Chapter 6). However, if there is no other choice, logic ground may have to act as a chassis point for one end of the cable shield. This is far from an ideal solution and is not certain to improve ESD immunity. In fact, it may sometimes cause increased ESD problems. The critical point is whether ESD radiation from an unterminated cable shield into various inputs will cause more, or less, problems than would the extra ESD noise coupled from the shield into logic ground via a capacitor. In general, the better the logic ground system, the greater the probability that this method will improve ESD immunity.

The required voltage rating for the shield coupling capacitor is directly related to the ratio of its capacitance to the capacitance of the ESD source. Remember, for a given charge level, the voltage level on an object is determined by the capacitance of the object. The coupling capacitance is determined by the frequencies to be coupled. A typical cable already has a chassis to logic ground capacitance of a few hundred picofarads. Therefore, it will accomplish little to add a few hundred more. In order to gain any significant advantage, the coupling capacitor from chassis to logic ground must be at least 1000 pF. On the other hand, a very large capacitor will have so much stray inductance that it won't couple high frequencies well. This means the coupling capacitor typically should be less than 0.01 μF. (Parallel combinations of smaller capacitors may be used to achieve better coupling, provided stray inductance of connections is kept small.)

A 3900 pF, 1 kV ceramic capacitor would be a reasonable choice for the coupling capacitor. This capacitor, in parallel with the existing cable capacitance, will couple a fairly wide range of ESD frequencies. What's more, a 20 kV charge on a 150 pF human body capacitance will be reduced to less than 1 kV when applied to the 3900 pF coupling capacitor.

This discussion of capacitance in cables leads to the subject of inductive filtering for cables (typically via use of ferrite beads). *Common mode* ferrite filters on otherwise well-designed shielded cables are not generally a good idea from an ESD standpoint. Since no shield is perfect, ESD current on a cable shield within a ferrite will induce opposing currents in lines within the shield. Of course, the same is true without a ferrite, but the ferrite increases the mutual inductance between shield and internal lines, and thus boosts the effect. If ferrite beads are required on cables (to reduce emissions for FCC, etc.), the following guideline should be remembered.

XI. The ESD Current Path Should Not Be Included in the Ferrite With Other Lines (It Is Best if the ESD Path Doesn't Flow Through Any Ferrites)

However, if the cable has no shield, or if the shield is not very effective, and the ESD noise is induced equally (common mode) into all lines within a cable, a common mode ferrite may be helpful. In such cases, all lines in the cable will have significant levels of induced interference, and the common mode ferrite will, of course, reduce this common mode interference. In these cases, use the following guideline.

XII. If a Ferrite Bead Is Used on Cable Signal Lines, It Will Be Most Effective when Placed at the Receiver End of the Signal Line, So It Can Filter Out Noise Picked Up on the Signal Line

There are times when cable cordage has more wires than are necessary. In these cases there are two possible ways to handle these extra lines.

XIII. Extra Lines in a Cable Must Be Either Clipped Off or Connected Properly:

A. Clipped off so they do not extend beyond the shield. (Don't leave a floating antenna to conduct noise inside the shield.)
B. Connected in parallel with other lines within the cable. (Don't connect in parallel with the shield, unless great care is taken to keep loop area of this parallel path to a minimum.)

XIV. Flat Cables Should Have a Logic Ground Line Beside Every Other Line, and Sensitive Signal Lines Should Be in the Center Conductors

Even a shielded flat cable will allow some flux through. Therefore, to keep the loop area to a minimum (see the discussion of loop area in Chapter 4) each cable line should be next to ground. Also, sensitive lines must be as far as possible from the cable edges, where the most leakage will occur.

Summary of Cable Design Guidelines

1. Electrical connections must use materials no more than 0.75 V apart from each other in the electrochemical series.

2. The anodic (more positive) material must have a larger uncoated surface area than the uncoated surface of the cathodic material.
3. Use shielded cables and connect the shield *only* to chassis ground. The arc path from chassis to other pins must be at least 2.2 mm.
4. The shield material must be at least 0.025 mm thick. (100 percent coverage is preferred.)
5. Connect the cable shield to chassis at both ends of the cable. A metallic connection is preferred, but a high frequency (capacitive) connection can be used if required to prevent a significant ground loop problem.
6. The cable shield must connect to the chassis within 4 cm of the cable entry/exit point, and the unshielded portion of the cable must be less than 4 cm.
7. Extra wires in the cable must be trimmed so they don't extend beyond the shield, or must be connected in parallel with the other lines.
8. A shield which is also a chassis ground path shouldn't normally go through a ferrite bead, and certainly not ferrite beads in common with other lines.
9. If ferrite beads are used, they should be at the receiver end of the cable.
10. If the shield can't be connected to chassis at one end of the cable, connect it to logic ground via a 3900 pF, 1 kV ceramic capacitor. The provision for this capacitor should allow it to be an option.

6 Enclosure Design Guidelines

Proper system enclosure design can increase electrostatic discharge (ESD) immunity both directly and indirectly. Enclosures can aid indirectly by allowing other portions of the design such as the cables or printed wiring boards (PWBs) to be done in a proper manner. The guidelines that follow relate to this indirect aid, and result in enclosure designs which will allow guidelines in other chapters to be carried out. PWB related enclosure guidelines are discussed first.

I. Metal Portions of Enclosures Should Be Connected to Chassis Ground (If this isn't possible, see III)

II. Portions of Enclosures That Are Metal, and Are Not Insulated, and Are Connected to Chassis Ground, Should Be Kept 2.2 mm From Exposed Electronic Components or Lines

III. Metal Portions of Enclosures That Aren't Connected to Chassis Ground and Are Not Insulated Must Be Kept At Least 2 cm Away From Exposed Electronic Components and Lines

IV. Avoid Sharp Edges on Metal Components Because They Encourage Secondary Arcs

The above guidelines are intended to prevent arcing inside enclosures. In air, a 20 kV voltage can arc about 2 cm. However, a relatively well-grounded chassis will typically not develop enough voltage to arc very far. The 2.2 mm guideline above refers to a small, sharp-edged metal object connected to chassis ground via an eleven foot cable shield. These guidelines are based on the fact that the dielectric separating electronic lines and components from chassis ground is typically air. If the electronics are completely sealed within a stronger dielectric material, such as plastic, then less separation is required. Remember, however, that plastic parts often have seams and openings which are not air tight. Be sure arcing can't occur via these openings.

The subject of seams and openings in plastic parts also relates to the subject of arcing directly from a person to the system electronics.

V. Design System Enclosures So an Operator Cannot Approach Closer Than 2 cm to Exposed Nongrounded Metal, Electronic Components, or Electronic Lines; *or* So a Chassis Grounded Object Is Between the Operator and the Electronics (Ungrounded Screws Are an Especially Troublesome Path For ESD Entry Into a System)

Even if a chassis grounded object intercepts the discharge before it can reach the electronics, it is best if the enclosure design provides the maximum separation from ESD source to ESD target. Although a discharge to chassis ground is not likely to damage equipment, it can still create strong near fields that must be dealt with. If no discharge is possible, these very strong fields won't exist. If nothing else, it is good to have a 5 mm separation from the person or other ESD source to possible chassis discharge points. That will prevent arcs at voltages up to above 5 kV.

It may sometimes be necessary to use a "crooked" path to increase the length of an arc path. For example, both of the seams shown in Figure 28 are in the same thickness of plastic, but one has a much longer arc path than the other.

Likewise, the examples of cooling slots, shown in Figure 29, illustrate how it is possible to get a longer arc path by adding an inner barrier.

In addition to the prevention of arcing, the enclosure design can help reduce PWB problems due to field effects on long lines and large loops.

VI. The Enclosures Should Not Constrict PWB Designs So Much That It Is Impossible to Include Sufficient Logic Ground Grid or Chassis Grounding

On a well-designed PWB, little extra room should be required to improve ESD immunity, but there will be some slight increase. If the design is too "tight" the first items to be sacrificed will probably be ESD prevention items.

VII. Enclosure Designs Should Allow the I/O Devices to Remain Close to the I/O Connector and Each Other

Don't force the PWB designer to run long PWB traces from the I/O connector to the electronics, or from component to component.

Fig. 28: *(left) Short arc path seam; (right) long arc path seam.*

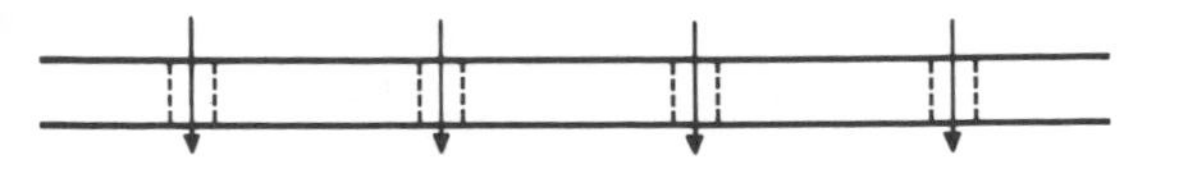

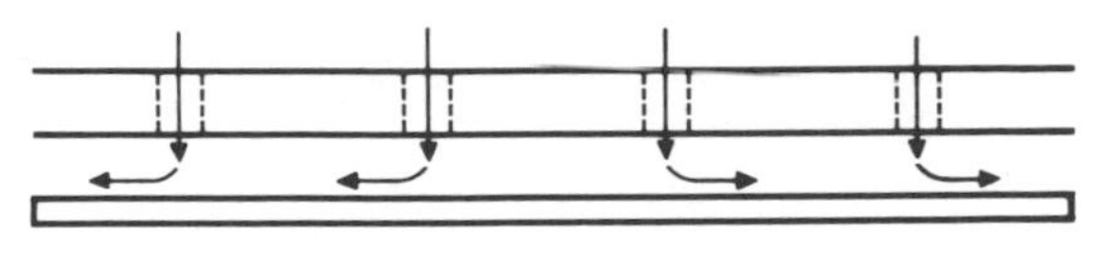

Fig. 29: *(left) Typical cooling slots; (right) cooling slots with arc barrier.*

VIII. Try to Locate the I/O Cable Entry Point in a Central Location on Each Enclosure

This will keep the signal and power supply line lengths as short as possible in all directions within the enclosures.

This leads naturally to those enclosure guidelines which enable the cable guidelines to be followed.

IX. There Must Be a Provision for a Short, Low Corrosion Connection of Each Cable Shield to the Chassis; This Connection Must Be Within 4 cm of the Cable Entry Point(s)

X. Connectors Must Be Within 4 cm of the Cable Entry and Chassis Connection Point(s), So the Unshielded Portion of the Cable Doesn't Exceed 4 cm

The subject of connection corrosion is covered in Chapter 5 and will not be repeated here, except to say that corrosion must be controlled. The 4 cm limitation is necessary to reduce the tendency of cable lines to act as antennas for ESD noise.

XI. If a Cable Has a Ferrite Added, this Ferrite Must Also Be Near the Cable Entry Point

The enclosure may have to be designed to accommodate the ferrite at the cable entry point.

Prior to the discussion of those enclosure guidelines which result directly in reduction of ESD problems, a short review of the ESD phenomena is helpful.

In other chapters, ESD problems that resulted when a person discharged directly to the electronics or to a chassis ground were discussed. The discharge resulted in direct charge injection, and this current generated E and H fields which caused field effects. Prior to the actual discharge an electrostatic field existed, and it also had potential effects. It should be pointed out that the fields, both static and dynamic, would exist even if the discharge was to an object near the system. (Of course, the charge injection effects would not exist in this case.)

The fact that fields exist for any nearby discharge is important! People often think they can solve all their ESD problems simply by designing their system enclosures so arcing to the equipment can't take place. Preventing arcs directly to the equipment does help, because these very near arcs have stronger fields. However, a discharge to the desk upon which a terminal is sitting will still have a potential for problems due to field effects. If the system enclosures contain no shielding, the only methods of dealing with these field effects are those discussed in other chapters, and those discussed previously in this chapter. In some cases those methods may be sufficient; however, it is usually not possible to be certain until the design is complete and tested. *For this reason, it is a good idea to design all enclosures so there is the possibility to add shielding.*

Because of the high frequency involved, the choice of shielding material for ESD is fairly flexible. Virtually any metal (copper, aluminum, steel, etc.) is acceptable. The only requirement is that the metal have some minimal thickness. A rule-of-thumb is that the thickness must be at least one skin depth at 75 MHz. (This would mean that aluminum and copper should be 0.025 mm and steel should be 0.05 mm.) Of course, a thicker shield is even better, but these minimum levels will typically provide acceptable results.

Although the resistance of a conductive coating or conductive plastic can be higher than metal, many conductive coatings and plastics will make acceptable ESD shields.

Since the choice of shielding material is not very limited from an ESD standpoint, the decision is based on manufacturing considerations. Presently, the basic choices are: full metal enclosure; plastic enclosure with foil (usually with a plastic backing), conductive painting, or plating; or conductive plastic. Each has advantages and disadvantages.

A full metal enclosure made of steel or aluminum may be an excellent shield for the electronics within. Unfortunately, such an enclosure is usually heavy, expensive, and not acceptable cosmetically. For this reason, most enclosures today are plastic. Shielding is then accomplished by one of the following four methods.

Foil usually places the fewest design limitations on the enclosure, but may be difficult to install in automated assembly lines. Also, care must be taken that the foil edges, or burrs, don't short to the PWB or components, and that the foil isn't torn or cracked. (Foil shields must be punched out so the edges bend away from the electronics.)

Conductive painting typically involves inexpensive equipment, but the enclosure must be designed so it is easy to spray all surfaces. Paints can flake or peel, and the part to be painted must be masked to prevent over-spray. In addition, if decorative paints are to be applied over the conductive paint, care must be taken that the decorative coat doesn't act as a solvent to disturb the conductive coating. Although easier to apply than foil, conductive painting is usually more labor-intensive than plating.

Plating requires expensive equipment, but for large quantities, it can be fairly inexpensive. The plating process must be set up so the required shield thickness is attained. Since plating usually results in a coating on both the inside and outside of the plastic, the thickness on each side need only be half as thick as the total required. Plating places minor constraints on the enclosure design (e.g., all corners must be slightly rounded), but it results in a very uniform shield that typically won't flake or peel. The biggest disadvantage of plating is that a decorative paint coating must be applied to cover it. This additional step negates some of the savings. (Single sided plating is possible, but this requires a process similar to painting prior to the plating.)

Conductive plastics require no additional steps to plate or conductively paint. However, conductive plastics can place limitations on mold design and can cause the molds to wear more rapidly. In addition, the conductive additives can degrade plastic parameters. As with other shielding methods, conductive plastics are more expensive than normal plastic.

If the electronics could be contained within a shielded enclosure that had no openings or holes, the choice of the shield material would determine the level of ESD problems. In fact, the choice of shield material usually is much less important than the physical design of the shield. Any hole or array of holes in a shield will act as a slot antenna. These slot antennas will act to couple signals that are outside, or on, the shield to the electronic circuits inside, where they can cause problems. The efficiency of this coupling depends on the length of the slot (or hole) compared to half of a wavelength of the external signal. If a slot is exactly one-half wavelength long, it is the most efficient. If a slot is exactly the same length as several half wavelengths, it is still efficient. Only when the slot is shorter than a half wavelength does it become less efficient. (The shorter it is, the less efficient it will be.)

As was mentioned in Chapter 2 discussing ESD effects, ESD noise contains a wide range of frequency components. Thus, the only way to ensure that a slot is not an efficient antenna at some ESD frequency is to make it much smaller than the wavelength of the highest ESD frequency. How small is small enough? In the field of electromagnetic interference (EMI) there are various rules-of-thumb. Some say a slot must be less than $\lambda/20$. Others say less than $\lambda/50$! Since the highest ESD frequency can be about 5 GHz, both of these rules-of-thumb would result in very severe constraints. The following is a more manageable rule-of-thumb:

XII. No Slot or Hole Should Have a *Long* Dimension Greater Than 2 cm

Note this refers to the *long* dimension of the slot! This means the largest possible hole in an enclosure is 2 cm × 2 cm! (This is roughly $\lambda/3$ at 5 GHz.) Since the $\lambda/20$ and $\lambda/50$ rules weren't used, more noise will get through a shield designed to meet the above guideline XII. This means that the designs

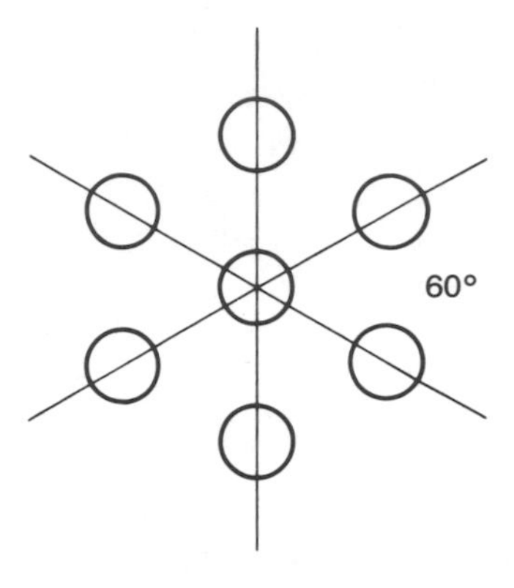

Fig. 30: *60° array of holes.*

of the firmware, cable, PWB, etc., are still important even when using a shield. (We are effectively using lower cost fixes to allow a reduction in shielding, which is typically more expensive. Of course, if there is no extra cost to make smaller shield openings, that is preferable.)

The primary methods of keeping slot antenna effects to a minimum are as follows:

XIII. Use Several Small Openings Instead of One Large Opening

XIV. The Space Between Openings Must Equal the Largest Dimension of the Opening

It is clear that the length of a slot will be reduced if it is made into several shorter slots. If several slots are close together, however, some of this advantage is lost because the slots are not truly independent. In order to be truly independent, the distance between the slots, or holes, must be equal to the long dimension of the hole. For maximum shielding and maximum hole area (as would be required for cooling vents), use a ''60 degree array'' of holes as shown in Figure 30. The distance from one hole to the next must be equal, or greater than, the hole diameter. The ''60 degree array'' has about 23 percent open (hole) area, no matter what the hole size. Of course, a smaller hole size results in better shielding.

In addition to hole size and relative placement, the placement of slots in the equipment shield can matter.

XV. Don't Place a Slot Near a Chassis Cable Ground Connection Point, or Near Sensitive Signal Lines or Devices

A slot antenna converts currents flowing around it into fields that it radiates. Therefore, it is a poor idea to place a slot in, or near, the system ground path for potential ESD currents. Also, to be sure noise leakage does not cause problems, slot antennas should not be near sensitive susceptors.

All the rules dealing with slots and holes also apply to seams in enclosures. This is because no seam is perfect. In a perfect seam, mating surfaces would be perfectly matched, free of corrosion, and they would remain that way. In short, there would be no apparent electrical opening. In reality, seams don't match perfectly, and they are never perfectly conductive because of corrosion. (Metals high in the electrochemical series corrode almost the instant they contact air, and cleaning has a limited effect.) Each gap in a shield seam is a slot antenna. Use the following methods to handle these slots.

XVI. Electrically Fasten the Shielding Seams at Several Points to Reduce the Slot Length That Can Exist (Screws, Clips, etc.)

If the fastener is a self-taping screw, be sure the burr created doesn't cause a gap. Also, tighten screws in the middle of the enclosure first, and work to the end to prevent buckling. When using screws, don't depend on the small surface area of the threads for good electrical contact. The purpose of the screw is to make the seam halves contact better. In the case of conductive plastic or of plated, or conductively painted, plastic; fingers, tabs, or clips can be designed in to connect across the seam.

XVII. If Necessary, Use Conductive Gaskets to Fill Gaps That Remain in Seams

Gaskets are usually used where it is not acceptable to use more fasteners, or where extra fasteners still don't result in a tight enough seam. However, some gaskets are very costly.

XVIII. Gaskets and Fasteners Both Should Be Chosen to Keep Corrosion to a Minimum (See the corrosion discussion in Chapter 5)

From an electrochemical standpoint, items with a thin plating (such as gaskets often have) act more like the base metal than the plating. One special method of reducing a seam slot is to cover it with foil tape.

XIX. If Foil Tape Is Used, It Must Make Electrical Contact With the Rest of the Shield

It was previously mentioned that no slot should be longer than 2 cm, and this means a simple butt seam should not have more than 2 cm between contact points. Clearly this places a burden on the enclosure designer. Fortunately, there is a way to reduce this burden. In order to limit the amount of noise passing through a seam, the seam should be designed to look like a waveguide beyond cut-off frequency. As a rule-of-thumb, if an open-ended (unterminated) waveguide has a length five times its diameter, it will drastically attenuate input signals.

It is almost always possible to make the overlap at least five times the seam gap width. This will result in high attentuation of noise signals that are polarized with their *E*-fields parallel to the seam. Unfortunately, for signals perpendicular to the seam, it is not so easy to create a cut-off waveguide effect because the overlap would have to be five times the distance between contact points. However, even an overlap that is equal to the distance between contact points will provide about 17 dB of attenuation for those signals with their *E*-fields polarized "across" the seam gap.

XX. A Shield Seam Should Be Overlapped, and the Overlap Should Be At Least Five Times the Gap Width, and Equal to the Distance Between Contact Points

Figure 31 shows examples of gap, overlapped seams, and contact points. (Note that the width of the gap in the examples of Figure 31 are unchanged,

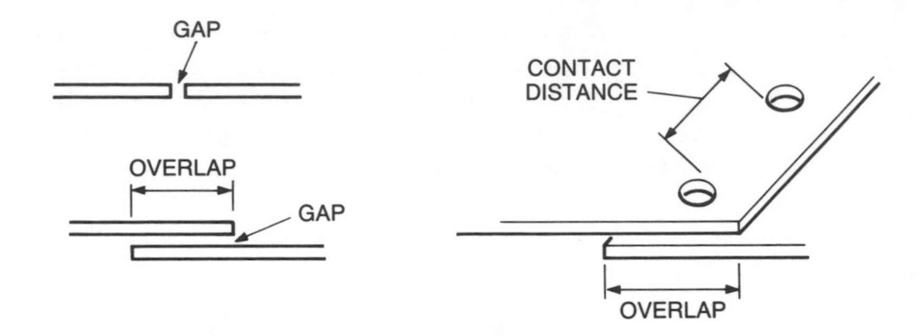

Fig. 31: *Overlap seams.*

but the overlapped seam will shield better.) Since the overlapped seam shields better, the fasteners may not need to be placed every 2 cm.

The concept of turning a hole into a waveguide is important! It can be applied not only to seams, but any shield opening. For example, a cooling vent can be made using a "honeycomb" construction as shown in Figure 32. Each "cell" of the honeycomb acts like a waveguide that will attenuate an applied signal. As before, the amount of attenuation depends on the dimensions of the waveguide. A large opening in each cell will require a longer waveguide section.

Instead of a honeycomb shape, other shapes can be used for openings and vents. The exact attenuation varies depending on the exact design, but use of the 5-to-1 length-to-opening ratio will usually result in fairly high attenuation. Figure 33 shows an alternate design for cooling openings in conductively painted plastic. In this case, the chassis grounded arc barrier is also used to create noise attenuation by appearing to be a waveguide beyond cutoff.

XXI. Bonding Straps Used to Connect Various Chassis, or Enclosure Sections, Must Be Kept Short and Kept Away From Sensitive Electronics. Bonding Straps Should Also Be Wide; It Is Recommended That Bonding Straps Be No More Than Five Times Longer Than They Are Wide

Bonding straps will radiate noise if ESD currents flow through them.

Summary of Enclosure Design Guidelines

The following summary details enclosure design requirements necessary to prevent ESD related problems:

1. The enclosure design must ensure that uninsulated electronic components and lines have at least a 2 cm arcing distance from ungrounded metal objects that may be touched by the operator.
2. The enclosure design must ensure that uninsulated electronic components and lines are at least 2.2 mm from any item connected to chassis ground.

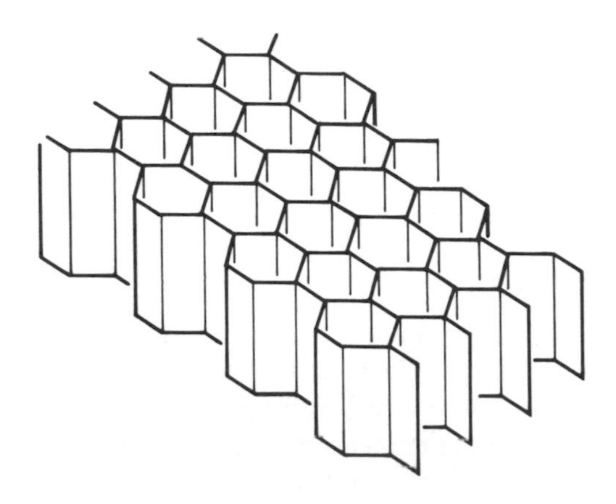

Fig. 32: *"Honeycomb" cooling vent.*

3. The enclosure design must ensure that uninsulated electronic components and lines are at least a 2 cm arcing distance from the operator, or that a chassis ground point is between the operator and electronics.
4. The enclosure design should allow the electronic devices to be grouped together. (If possible, the I/O connector should be centrally located.)
5. The enclosure design must allow sufficient room for the PWB, so PWB design guidelines can be followed.
6. All shield materials must have an EMF within 0.75 V (in the electrochemical series) of the metal they connect to. If not, an intermediate metal connection device must be used.
7. All designs must make provision for the addition of shielding.

The following specifications refer to shielding design.

8. No slot seam or hole in the shield may have an opening dimension greater than 2 cm, unless the length-to-opening ratio of the hole is at least 5 to 1 (see Figure 34).
9. The two acceptable holes on the left of Figure 34 may also represent seams. This means a seam must have an overlap at least five times the gap width.
10. If requirement 9 can't be met for shield seams, use gaskets to fill the seam gaps, or use fasteners every 2 cm along the seam.

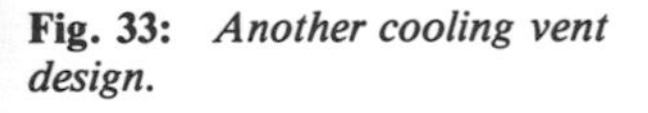

Fig. 33: *Another cooling vent design.*

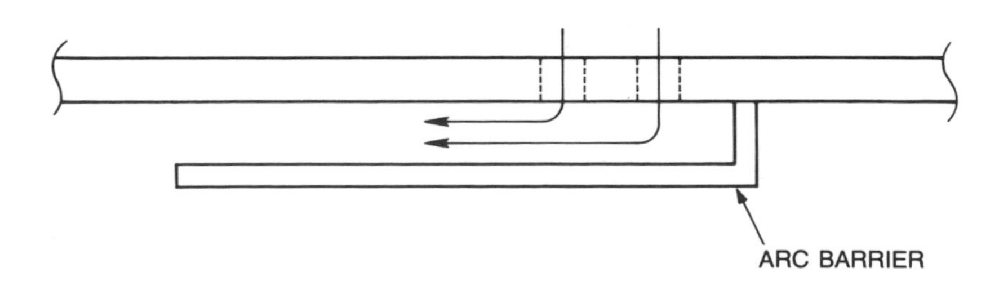

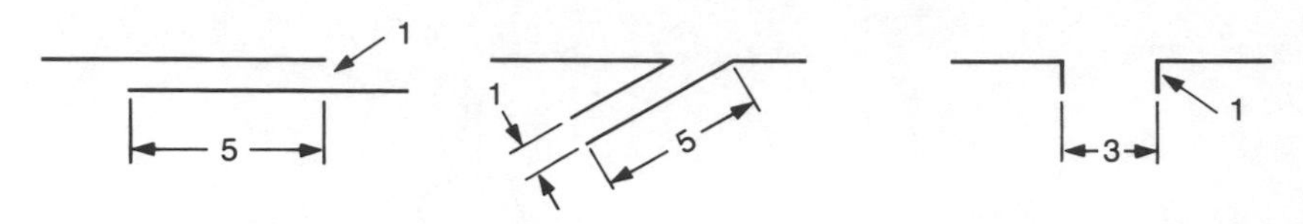

Fig. 34: *(left) These holes are okay; (right) hole not acceptable if greater than 2 cm.*

11. If several holes are required, the space between holes should equal the largest diameter of the hole.
12. Use several small openings instead of one large opening. (Consider using the 60 degree hole array of Figure 30.)
13. Do not place a shield hole near a point where the shield connects to chassis ground, or near sensitive devices or lines.
14. If foil tape is used, it must make electrical contact with the shield.
15. Keep bonding straps short and wide.

7 Electronic Circuit Design Guidelines

As indicated in other chapters, electrostatic discharge (ESD) generated noise can enter the electronic system via either conduction or radiation. The methods detailed in the following electronic circuit guidelines are effective in coping with conducted ESD noise, including conducted noise that is induced by ESD radiation.

First, two general guidelines should be mentioned. They are as follows:

I. Circuit Designs Must Not Push Components to Their Limits

If a component is slightly damaged by ESD during manufacturing or in the field, it will often continue to operate indefinitely provided that the operation is not already stressful. Likewise, a part that isn't already under stress will be less likely to be damaged by the addition of ESD stress. It is difficult to put exact limits on the derating of components. There is little public information to prove that derating components by *Y* percent will reduce ESD failures by *X* percent. Clearly, however, no component should be forced to operate at levels beyond those well defined by the vendor's specification. For example, an LSTTL output, rated at 8 mA for 0.4 V, will accept much more than 8 mA at higher voltages. A designer may be tempted to use it to drive an LED (at 20 mA) rather than add a higher current driver IC. While it is true that the LSTTL IC can drive the LED, it will be working very hard to do so. The addition of ESD-caused latent damage or induced currents may easily push it to complete failure.

II. No Design Should Result in a Circuit or System that Can Disable Itself for an Indefinite Period

If a circuit has a disabled state, you may be certain that ESD will eventually force that state. There must be some means to undo this condition, even if other portions of the system are not aware it has happened. This applies at both the circuit level and the system level. For example, if a keyboard can be disabled, or forced into a wait condition, something should periodically check the keyboard to ensure it hasn't been stopped by static. Likewise, an IC that can be forced to a low current "sleep" state should be awakened periodically to ensure there is no ESD-caused lockup.

In addition to hardware microprocessor lockup, firmware lockups may occur in which the processor is unable to restore operation due to endless program loops, etc. People often speak of using "watchdog" circuits to prevent such firmware lockups. Such circuits may indeed be required in specific cases; however, well-written firmware usually will not have lockup problems (see Chapter 3).

As originally noted, electronic circuit design guidelines deal with conducted ESD noise. These guidelines fall into the two categories:

1. Preventing the noise from being conducted to sensitive inputs.
2. Dealing with noise after it is conducted to sensitive inputs.

Category 2 includes fewer options, so it will be discussed first.

III. Double-check Critical Inputs

(This is mandatory on all designs.) An ESD pulse typically will not last longer than a few microseconds. Therefore, any input which can cause an irreversible action must be checked at least twice, and the checks must be several microseconds apart. This will insure that a single ESD pulse can't cause problems.

In order to prevent problems from the multiple ESD discharges that can occur in a single ESD event, parity, and frame error checks are helpful. Such checks will reduce the probability that an incorrect code can cause problems.

IV. Use Less Sensitive and Slower (Narrower Bandwidth) Devices (This may not always be possible)

For example, some early PROMs are clearly more sensitive than those developed more recently. Also, the type of input matters.

Two specific methods of reducing the sensitivity of inputs are as follows:

A. Use differential I/O schemes, such as RS422. Differential I/O may not help as much as may be expected. It will only be successful if steps are taken to insure that the ESD noise is *identical* (common mode) on the differential I/O lines. This relates to PWB, cabling, and enclosure requirements. If PWBs, cables, and enclosures aren't designed properly, the differential I/O may provide little aid. On the other hand, if the cables, PWBs, and enclosures are well designed, there may be little left for a differential I/O to accomplish.
B. Don't use edge-triggered or wide bandwidth logic. Edge-triggered devices are glitch amplifiers. If a glitch has a high amplitude, it can trigger a logic device even when it is much narrower than the propagation delay of the device. However, once the device is triggered, the effect will travel through the system circuitry as if a real signal had caused it. Likewise, wide bandwidth devices will propagate very narrow glitches.

Unfortunately, it is almost impossible to design without some edge-triggered logic. This brings us back to those guidelines related to category one, which details how to prevent noise from being conducted into sensitive inputs. There are four ways to prevent noise from being conducted to an input discussed in guidelines V, VI, and VII, and finally in guideline VIII.

V. Disconnect Sensitive Inputs From Those Lines Which Have ESD Noise (Disconnect the Antenna)

Obviously, you can't disconnect all lines from a sensitive input because it will no longer function. However, designers often design in lines which will almost certainly cause ESD problems. For example, it is not a good idea for the system RESET line to be connected from a terminal to the keyboard via a six foot cable. Induced noise on such a long RESET line could cause frequent system reset, which would be unacceptable.

VI. Connect Floating Inputs High or Low

A floating input is typically resting at its switching threshold. A small noise signal picked up and conducted to the input by the component lead of such a device can easily be enough to exceed the switching threshold and cause an output. Even if the output isn't used, the device could experience a short period of oscillation, which could result in noise on the power and ground lines. (Note that some device inputs have built in pull-ups, etc., to insure they are not actually floating.)

VII. Filter Sensitive Inputs

As was mentioned in guideline V, even sensitive devices must have at least one input signal. Filtering can protect such inputs. Filters for ESD noise must be as close as possible to the receptor to insure they are most effective. A filter placed five centimeters away from an input leaves five centimeters of antenna connected to the input. (This, of course, assumes that the input line is not shielded.) This requirement, that ESD filters be close to the receptor, can result in a substantial problem. It is usually not cost effective to filter every single input in a given design. Typically, therefore, filtering is only applied to those inputs that are especially sensitive, or which may be in contact with higher levels of ESD noise. For example, I/O devices may see higher noise levels because the long cables to which they are attached act as good antennas.

Fortunately, if the PWB, firmware, cable, and enclosure design are well done, filtering for ESD noise may be not be required. This means that, if possible, filtering should be designed in as an option. The designer should depend primarily on firmware, PWB, cable, and enclosure design to prevent ESD problems. Filtering can be added if the ESD immunity is unacceptable and other solutions aren't possible.

If filters are used for ESD, as suggested in guideline VII above, they may be formed by the addition of shunt capacitance or series inductance. (Of course, both filtering methods may be combined.) First, capacitive filtering will be covered.

If capacitors were ideal, the proper value to use would be "as big as physically possible." Unfortunately, real capacitors have stray inductance, and this inductance generally increases as capacitance increases. At high frequencies, the inductive effect predominates, and the capacitor is a very poor shunt filter. In order to have a low impedance at high frequency, the capacitor must have a low inductance, and thus, low capacitance.

To determine what value of capacitor is required, consider the "antenna" you are dealing with. The input to a device has to be filtered because it is connected to an antenna, which picks up the fields radiated by ESD currents. The "input antenna" will only be efficient if it is at least a significant fraction of a wavelength in length. This means the lowest frequency to be filtered will depend on the size of the antenna to which the input is connected. Lines on a well designed PWB typically will be efficient antennas for frequencies above 100 MHz. At these frequencies, even ceramic filter capacitors could have values no more than tens of picofarads. Larger capacitors would have too much stray inductance. However, a well designed PWB, in which signal lines are kept close to ground, will already have stray capacitance amounting to tens of picofarads. *This means the addition of shunt capacitance filtering on signal lines of well-designed PWBs typically will not be much help. It is far better to make use of the shunt capacitance obtained by proper PWB layout.*

Cables between equipment can easily be efficient antennas at a frequency of 10 MHz and up. For frequencies in tens of megahertz, ceramic capacitors could have values of a few hundred picofarads, and still have reasonably low stray inductance. However, the stray capacitance from one line to another, within a cable, can also amount to a few hundred picofarads. (This is especially true if the cable is shielded.) *This means that the effect of added shunt capacitance will be limited even when filtering inputs from cables. A well designed cable will do more to reduce ESD problems than a shunt capacitance filter.*

This doesn't mean that shunt capacitors will not help at all, but it does mean that the degree of help will be limited. If a design is close to acceptance levels and you're grasping at straws, try adding shunt capacitors. They may be just enough help to get you through. However, if shunt capacitors are used, heed the following guidelines:

A. Shunt capacitors should be placed close to the input, and should be designed in as options. It was previously mentioned that ESD filtering must be done close to the inputs to be truly effective. However, designers who are accustomed to filtering outputs to pass emission regulations may completely forget to filter the inputs for ESD.

B. Shunt capacitors must work at the proper frequencies. In addition to stray inductance, other physical aspects of the capacitor can affect its operation at high frequencies. Figure out the length of the antenna connected to the input to be filtered. Use that antenna length to calculate the frequency for which the antenna is at least 1/8 wavelength long ($f = 3 \times 10^8$ m/[8 × antenna length in meters]). This formula is not precisely correct because it is for a single wire with air dielectric. However, it is good enough for this purpose. The capacitor you chose must operate *well* above this frequency.

C. Connect the shunt capacitance between the signal line and logic ground line. Although the shunt capacitor could be connected to chassis ground, it can be difficult to make such a connection successfully. Therefore, you should not normally shunt to chassis ground.

The above guideline C is directly opposed to what you will often read for normal EMI filtering. The recommendation to connect filter capacitors to chassis ground is based on the assumption that chassis ground paths are lower impedance than logic ground paths. While this is normally true, if the PWB and cable guidelines are followed, logic ground impedances can also be kept very small. More important, there is a significant difference between the ESD problem and EMI problems. Specifically, ESD can *directly* inject a significant current into the discharge target. On a well designed system, the only possible discharge target should be chassis ground. If the capacitive coupling betweeen chassis ground and the signal lines is increased, *more*, not less, noise may be induced into the signal lines from chassis ground.

In order for chassis connection to work well, chassis ground must be capacitively well coupled to the entire electronic circuit. This includes not only all signal lines, but logic ground and power lines. In this case, noise on chassis ground would be felt equally by all other lines (will be common mode), and thus no differential signals could result.

However, it is difficult to implement such a design using individual shunt capacitors. This is because of the frequency limitations already mentioned, and the problem of truly coupling equally to all portions of the circuit. This approach is easier to carry out via cable, PWB, and enclosure solutions, than by discrete capacitive shunting methods.

If you aren't discouraged enough at the prospect of trying to use shunt capacitance filters, there is one more point to note. Additional shunt capacitance will increase "ringing" on signal lines. The ringing is caused by the line inductance and capacitance which forms a tank circuit. Figure 35 should make it clear that an increase in either shunt capacitance or series inductance will increase ringing.

This deals a serious blow to the other ESD filtering scheme, which is the addition of series inductance. Not only that, but the problem of stray capacitance in inductors is mirrored by the same types of problems that stray inductance may cause in shunt capacitors. This means no single inductor is capable of acting as a filter over the entire range of ESD frequencies.

Fortunately, there is a very special type of impedance device that

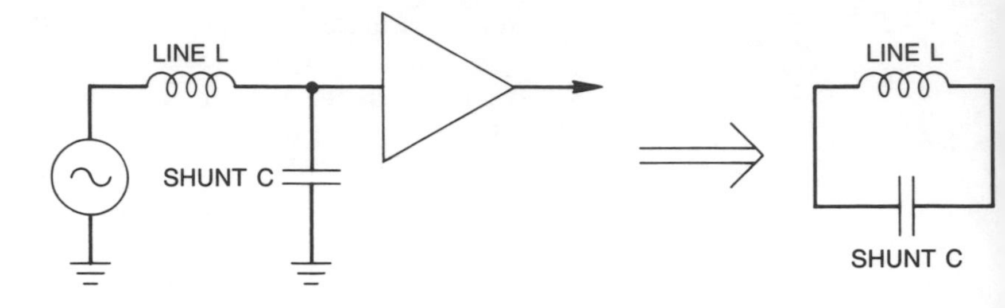

Fig. 35: *Signal lines form an equivalent "tank" circuit.*

overcomes the problems of a normal inductor. That is the ferrite "bead." At low frequencies, the ferrite bead acts like an inductor. At high frequencies, the bead becomes resistive. Therefore, it acts as a series impedance over a broad range of frequencies. This makes the ferrite much more useful than either a capacitor or an inductor. However, as was the case with shunt capacitance filters, a well designed PWB, cable, and enclosure can often reduce ESD problems to such a degree that even ferrite filtering may provide little improvement. This leads to the next guideline.

D. Design in ferrite ESD filters as an option. If the ferrites aren't helping don't install them.

E. The ferrite ESD filter must be close to the input it is filtering as has been discussed earlier. Ferrites should have *nothing* between them and the input they are filtering.

F. Each line to be ESD filtered should have an independent ferrite bead. Don't use a common mode filter unless it is certain that ESD will induce common mode interference. It is a popular practice to encircle an entire shielded cable with one large ferrite bead. This practice may help reduce RFI emissions from the shield, but will often increase ESD problems. It is clear that a ferrite bead on a single line will resist an induced current no matter what the direction of the current flow. If two lines are enclosed by the same ferrite, this is not always the case. If the current induced in both lines is in the same direction and inphase (is common mode), then the ferrite will indeed resist it. However, if the current induced in one line is opposite to the current induced in the other line, the inductive effects of the two currents in the ferrite cancel, and neither current flow is impeded. What is more, if one of the induced currents is stronger than the other, the effect of the two currents in the ferrite will tend to reduce the stronger current, but increase the weaker current! Therefore, a strong ESD current flowing in one line will tend to induce an opposing noise current in other lines, and the common mode ferrite will boost this effect! Increased induced currents on signal lines obviously will increase ESD-related problems.

G. Generally, it is better not to use multiturn ferrites for ESD filtering. A ferrite, designed to allow the wire to be wrapped around it, will have a larger inductance. Unfortunately, it will also have a larger stray capacitance. This creates the same problems that stray capacitance causes in normal inductors in that the ESD noise will be coupled around the ferrite by stray capacitance. If you want more ferrite effect, use longer beads, or more beads strung on the same straight wire.

H. Use the proper ferrite material for the ESD frequency range. Because of the broad working frequency range of ferrites, it is less difficult to select a proper ferrite than inductor. However, it is still necessary to be sure that the ferrite covers as much of the ESD frequence range (approximately 500 kHz to 5 GHz) as possible. The frequencies of 10 MHz to 1 GHz are especially important.

I. Don't let ferrites touch each other, or other circuit lines. Many ferrite materials have a very high resistance, and can be thought of as insulators. Some ferrite materials, however, have resistances on the order of 100 Kohm/cm^3. If they touch each other, or other circuits, cross-talk could result. Therefore, it is good practice to treat ferrites as conductive objects, and maintain isolation from other circuits. This also means that PWB lines shouldn't be routed under ferrites.

There is one other method to terminate input lines in order to protect sensitive inputs.

VIII. Protect Inputs with Very High Speed, High Current, Clamping Suppressors

A clamping suppressor will act to shunt off ESD noise above a certain level. This will obviously help reduce damage to sensitive inputs, and it will also somewhat reduce noise-related problems. As was previously mentioned, a high amplitude spike can trigger a device even if its duration is less than the propagation delay of the sensitive device. Since the suppressor will act to reduce the amplitude of the spike, it will also reduce the chance that the ESD-caused spike will trigger the electronics. Many ICs have built-in overvoltage protection diodes. These will also help reduce ESD problems, although they are not designed to absorb high currents.

Unfortunately, compared to ESD, even high speed transient suppressors are slow to turn on and clamp the ESD signal. Therefore, for the initial portion of the ESD signal, the suppressor will merely be a capacitive element. As a result, when installing suppressors, use the same guidelines as those used for shunt capacitors (guideline VII; A, B, and C, above). As with shunt capacitors, these suppressors will normally provide limited help against ESD, if the PWB, cable, and enclosure are already properly done.

However, there is one special case in which ferrites and clamping suppressors may provide significant help. If signal or power lines must be directly accessible (for example, the open pins of an unused connector), then a series ferrite combined with a shunt suppressor may be used to protect against damage due to direct discharges to the signal or power lines. Because the risk of damage is very great in this case, it may be necessary to shunt to chassis in order to keep the ESD noise common mode, and to provide the lowest possible shunt path impedance. However, by shunting to chassis designers should recognize that they may be improving immunity to damage at the expense of immunity to upset caused by discharges to chassis.

One other termination method is also possible, but is generally not helpful for ESD. That is the use of termination resistors to match the characteristic impedance of the line. In Chapter 2, it was pointed out that a change in impedance would only trade *E*-field for *H*-field effects, and vice versa. Therefore, in typical circuits, impedance matching solves one ESD problem at the expense of another. Impedance matching may be required for other reasons, but it is not generally a cure for ESD problems.

In concluding this discussion of electronic design guidelines, it should be apparent that only marginal improvement can be normally expected by following the filtering guidelines. The other guidelines in Chapters 3, 4, 5 and 6 concerning firmware, PWB, cable, and enclosure, respectively, are far more important from an ESD standpoint.

Summary of Electronic Circuit Design Guidelines

The following are ESD-related design practices that should be followed whenever possible. Items 1 through 3 are especially important and should be considered mandatory requirements.

1. No design should force a component to operate outside those levels which are well specified by the vendor.
2. No design should result in a circuit that can disable itself indefinitely.
3. All inputs should be double sampled, and the samples should be several microseconds apart.
4. Use parity and frame error checking whenever possible.
5. Use less sensitive and slower devices.
6. Don't design in extra ESD antennas, such as long reset lines.
7. Connect all floating inputs either high or low.
8. If ferrites are used for ESD problem prevention, follow these guidelines:
   a. Put ferrites next to the "input" they are filtering (within 2.5 cm). (Ferrites for ESD should be on inputs, not outputs.) If the electronic components are shielded, then the input is the entry point into the shielded region.
   b. No other component (except a connector) should exist between the ferrite and the input it is filtering.
   c. Each input to be filtered normally should have its own ferrite bead. Common mode filtering normally is not good for ESD protection.

d. Use the proper ferrite material.
e. Be wary of multiturn ferrites.
f. Don't let ferrites touch each other, or other PWB lines, or ground grids.
g. Design in ferrites as an option only.

8 Manufacturing, Shipping, and Installation Guidelines

Even the best system grounding path, shield, etc., will have no impact on electrostatic discharge (ESD) until after installation. Thus, on well designed products, the majority of actual ESD damage will often be done during manufacturing, shipping, and installation. Repairing a system after assembly is bad enough, but it is even worse if the failure doesn't surface until the system is installed.

Unfortunately, there is a tendency for personnel to assume that ESD damage can be found in the final test. This is not always true. It is quite common for a component (typically an IC) to be damaged, but not destroyed. A component with such a ''latent failure'' can then fail under the stress of continuous operation at the customer site. There is also a tendency for personnel to assume that ESD isn't present if they can't feel it. This is also not always true.

Normally, people can't feel ESD levels below 2 to 3 kV, but 2 kV is certainly sufficient to damage many items. What's more, there need not be a spark to destroy components. The electrostatic field itself can cause dielectric breakdown, even if an arc never occurs. To make matters worse, it is these imperceptible static problems that are those most likely to cause the latent failures that can be a prime cause of expensive field failures.

Because of the ''time bomb'' effect of latent failures, one can't wait for problems to occur before instituting ESD protection measures. Static protection must be continuously maintained. In order to understand static protection guidelines, it is necessary to review the basic ESD process outlined by the following five causes and effects.

- Charge is generated by bringing materials into contact.
- A charged dielectric holds charge and creates an electrostatic field.
- If the charged dielectric is near a conductor, the field will induce a charge in the conductor.
- Two conductors with significantly different voltage levels can break down the insulation dielectric (air, plastic, metal oxide, etc.) that separates them, and an arc will form to equalize the charges on the conductors.
- Both the dielectric breakdown and the arc current can result in permanant damage or destruction.

This list of cause and effect in the ESD process suggests several solutions

I. Keep Dielectric Materials to a Minimum

In today's world, we are surrounded by many highly dielectric materials and it is impossible to eliminate them totally. They must, however, be limited if there is to be any chance of coping with ESD. That is, the generation of charged objects must not be so rapid and numerous that they can't be discharged fast enough.

Examples of dielectrics that should be eliminated as much as possible are non-antistatic work surfaces, conveyors, carts (remember the wheels), boxes, totes, bags, and packing material. If especially static-sensitive devices are handled, floors, chairs, stools, personal clothing, and shoes should be limited to poor charge generators such as cotton and leather (not polyester and rubber).

II. When Dielectric Materials are Required, Keep Contact Between Them to a Minimum

Try not to store plastic parts in plastic boxes or bags. Try not to set nonconductive boxes on nonconductive work surfaces. Of course, if a dielectric item is not to touch another dielectric item, it must, by default, be in contact with a somewhat conductive (so-called static dissipative) material.

III. Conductive Materials Must Be Connected to Ground, If At All Possible (Remember, the human body is conductive)

A truly "floating" conductor can itself be charged by a dielectric. Grounding the conductor prevents this charge storage.

One point should be clearly understood. Just because a dielectric is touching grounded metal, that doesn't mean guideline II can be ignored. Grounded metal will not prevent a dielectric from being charged. This means the clothing of even a grounded person may become charged, and should still, therefore, be made of material which is a poor charge generator. Remember, since a dielectric is an insulator, there is no requirement for a charge to be equal in all portions. This explains why a dielectric cannot be quickly discharged by grounding it. In order to discharge a dielectric, a neutralizing charge must be applied over the whole surface (e.g., with an ionizer; see guideline IV below).

This is an appropriate point to comment on the practice of grounding personnel by having them touch a grounded conductive surface. This works only as long as they continue to touch the grounded surface. As soon as they let go, charges can again rapidly build up.

This is also an appropriate point to comment on the practice of putting a foil-backed tape on sensitive electronic components. The goal, of course, is to short all pins of the component so no voltage differentials can develop and cause damage.

Unfortunately, it is forgotten that the tape must eventually be removed. During tape removal, the dielectric of the tape is usually charged to a substantial level. The foil backing, of course, has an equally large induced charge. This means the component pins still touching the foil during removal will have a high charge, but the pins not touching the foil won't. Therefore, the tape creates the very problem it was to correct. The removal of the tape insures the component will be subjected to high voltage differentials on its pins. This would be prevented by following guidelines I and II. (Avoid dielectrics and avoid dielectric contact.)

IV. If It Is Not Possible, or Cost Effective, to Prevent Charge Buildup Completely, then Use Ionizers, or Other Devices, to Discharge Components Prior to Assembly

This guideline shouldn't be an excuse to forget guidelines I through III. Although a dielectric part can be discharged, it is better to prevent charge buildup. Remember that a charged dielectric creates an electrostatic field, and this field can damage components in less time than it takes to discharge the dielectric item. The philosophy normally should be to use such items as ionizers, etc., to prevent significant charging, not to bleed off charge that has been allowed to develop. (In this regard, it should be noted that ionizers which don't generate balanced charges may themselves be a source of charging.)

However, in spite of your best efforts, there will always be some charge buildup and, thus, there will always be electrostatic fields to contend with.

V. Minimize Handling of Components or Assemblies; Especially Handling Resulting in Conductive Contact

This applies especially to handling at places other than work stations, where static is controlled. However, even at static controlled workstations, it is good practice not to handle items more than necessary, and when handling items to touch only nonconductive portions (e.g., handle PWBs by the edges or extractors). When components or assemblies touch each other, or other surfaces, it is also good practice to keep sliding motions to a minimum, and not to touch conductive portions together unless necessary. Even when the person handling components or assemblies is grounded, the components or assemblies may themselves become charged by rubbing action. Then, if conductive portions contact other conductors, ESD may occur. (This is especially likely during installation.) Because the series resistances and inductances involved in such ESD may be much smaller than for human ESD, very large peak currents are often associated with this component/assembly ESD.

A common method by which assemblies may be charged is the use of a rubber eraser to clean connector contacts. Not only will an eraser often wear the gold plating off the contacts, but it may generate a significant charge in the assembly, due to the rubbing action. If contacts must be cleaned, use wipes treated with an antistatic cleaner.

One final point must be made about handling. A prime source of potential ESD problems is the supervisor who wanders along the production line examining products. Supervisors must follow the same ESD reduction guidelines that they require their personnel to follow.

By following the above handling guidelines, damage will be reduced. However, some components are especially sensitive, and will require additional protection. In a well controlled manufacturing environment, any component that may be damaged by 500 V or less should have additional protection, but in the less controlled shipping and installation environment, a component that may be damaged by 4000 V or less should have extra protection. This extra protection is shielding.

VI. Shield Especially Sensitive Electronic Components and Assemblies

Typically, shielding during the assembly, shipping, and installation process will be provided by conductive bags, totes, etc. Components or assemblies requiring shielding should be kept in the shield until it's necessary to remove them. It's also best not to transfer them from container to container.

If shielding is required, there are three requirements for these items.

A. The shield must be complete. For example, an open box will shield only from the sides and bottom, not from the top. Small holes in the shield, such as the holes on the ends of IC tubes, usually will not be a problem, but large unshielded areas severely reduce the shield effectiveness.

As was previously stated, conductive items must be grounded if they are to bleed charge off of other conductive items. This is also true of shields. A charged item, placed inside a floating shield, will charge the shield. However, the main purpose of a shield isn't to bleed off charge (although that is nice), but to protect the electronics inside from the effects of electrostatic fields caused by outside charge. As long as a shield *completely surrounds* the item it is protecting, it will be an effective shield whether it is connected to ground or not. A ground connection doesn't improve shielding, but by bleeding off charge, a ground connection will, in its own way, reduce electrostatic field effects. Therefore, it is a good idea to place bags, totes etc., on a grounded surface for at least one second prior to opening them. Also, it is a good idea to touch the shielded container prior to touching items inside.

B. The conductivity of the shield must be high enough to respond rapidly to field changes. This is because, in the most general case, shielding material reflects and absorbs unwanted *E* and *H* field energy. However, most ESD shielding products intended for manufacturing, shipping, and installation environments are primarily reflective shields. Reflection is related to the conductivity of the material. A shield reflects fields by reorienting its charge to oppose the external field. If the shield is a poor conductor, this charge realignment will take too long and damage will be done before shielding occurs.

The question of shielding is complicated, however, by the requirement that arcing also be prevented. For example, if an electronic PWB assembly were to become charged, the act of placing it in a highly conductive (e.g., metal) box would probably result in a discharge from the uncharged box to the charged PWB assembly. Such an arc would very likely result in damaged components.

C. Shielding used to protect items during transport should have a fairly conductive layer surrounded by a layer of much lower conductivity. The fairly conductive layer will respond rapidly to changes in external fields. The low conductivity layer will act to prevent arcing, if it comes into contact with charged conductors, and it will not itself generate or hold charge. As a rule of thumb, the fairly conductive layer should have a resistivity of less than 10,000 ohms per square. The low conductive layer should be about 10^{6}–10^{10} ohms per square.

This requirement in C above, that there be no low resistance arcing path, means a resistance guideline must be added related to grounding guidelines.

VII. The Ground Path for Conductive Items Must Have a Resistance of at Least 1 MΩ, but No More Than 100 MΩ

It was previously stated that all conductive items (including the human body) should be grounded, if possible. The above resistance rule should apply to all these grounded conductive items. The lower limit of 1 MΩ ensures there won't be a tendency of high current arcs to grounded objects. The upper limit of 100 MΩ ensures that any charge applied to the conductive item will bleed off quickly. When grounding personnel, or for any ground surface that people may touch, the ground resistance must be sufficient to prevent the flow of more than 0.5 mA. Larger currents can be dangerous to the human system. (The resistance in the ground path should be as close to that of the person as possible, to protect the person from inadvertent shorts to the ground strap. Also, personnel ground connections should be clearly different than AC power connections.)

The final two guidelines are relate not to technical problems, but to "holes" that commonly occur in a static protection system.

VIII. Static Prevention Must Take Place at *All* Points in the Manufacturing, Shipping, and Installation Process

Some companies only worry about static prevention on those processes after final test. They think they can "weed out" all ESD failures at final test, so they don't worry about processes prior to the final test; these people don't understand latent failures. Other companies are very careful right from the start until they enclose the unit in a plastic enclosure. After enclosure assembly, they figure the equipment is "insulated" so no failures are possible; these people don't understand that insulators are what generates charge. They also don't understand that discharge isn't the only failure mechanism, but that electrostatic fields can also cause failures. Most companies are very concerned about protecting electronic components, but they don't think about preventing damage due to charge buildup on plastic parts. They'll take their carefully protected PWB assemblies and install them in plastic enclosures that are charged to 5 kV. These people don't understand that electrostatic fields may be associated with insulators as well as conductors. And of course there are customers or installers who receive a carefully protected and packaged unit, which they carefully unpack and carry across a nylon carpet. Then, when they plug it in, a virtual lightning bolt jumps between the connector pins. The message should be clear: Every process, every person, and every part, requires protection, and this protection is required until the unit is installed and/or the "designed in" protection features become operative.

IX. Make Regular Checks to Ensure the ESD Protection System is Working Properly

A good ESD protection system can degrade if it is not maintained.

A. Make sure people understand and are following the guidelines. All the wrist straps and shields in the world won't help if they aren't used.

B. Make sure the equipment works. Check ground connections and surface conductivities. Be sure they are within spec (there are meters designed to check ESD protection devices). In a well maintained grounding system, voltages shouldn't exceed 100 volts. Ground straps, work surfaces, and shielding items all degrade with age, and should be replaced when they're no longer acceptable.

C. Make sure the procedures and equipment are effective by measuring charge levels to verify that charge is being controlled. In addition, an AM radio or other device can be used to detect sparks that are too small to feel.

By using these methods to "hunt" for static, you may find overlooked problem areas. Considering the cost associated with ESD problems, such checks should be frequent, perhaps even daily. Just because you think everything is okay doesn't mean it is. Constantly prove to yourself that the procedures and equipment are doing their job. Don't wait for the latent failure "time bomb" to go off.

Summary of Manufacturing, Shipping, and Installation Guidelines

The following steps are required in manufacturing, shipping, and installation in order to prevent ESD problems:

1. All work surfaces must be static dissipative (surface resistivity 10^6–10^9 ohms per square) and must be connected to ground (see 3).
2. All persons who handle unshielded components or assemblies must be wearing a wrist and/or heel strap which is connected to ground (see 3). (This includes supervisory personnel.)
3. All ground connections must have a resistance of between 10^6–10^8 ohms.
4. All dielectric items (nylon coats, plastic lunch boxes, etc.) that are not necessary for the manufacturing process must be stored, and kept, at least six feet away from the assembly line and work areas.
5. Dielectric (nonconductive) items which are required for the manufacturing process should rest directly on a conductive surface and should not touch another dielectric item. This applies for both work areas and transport.
6. Totes, bags, boxes, carts, conveyors, packing material, and other things (including people) used to transport, ship, and store components and assemblies either must be continuously grounded (see 3), or must be placed in contact with ground, or a grounded surface, for at least one second before components or assemblies are put in or taken out.
7. Keep handling to a minimum. Touch components and assemblies on nonconductive surfaces, and prevent conductive contact with other assemblies or surfaces whenever possible.
8. Components and assemblies designated as "especially ESD sensitive" must be kept in shielded containers (see 9) until they are installed. If these especially sensitive items are shipped as components, or bare assemblies, then they should be shipped in shielded packaging. (Generally, components that can be damaged by 500 V or less should be considered especially sensitive in a manufacturing environment, and components that can be damaged by 4000 V or less should be considered especially sensitive in a shipping or installation environment.)
9. If a shielded container is required, it must completely surround the

component or assembly (e.g., boxes must have lids). Any surface of a shielded container that contacts the components or assemblies must have a resistivity of 10^6–10^{10} ohms per square. However, the shield layer must have a resistivity of less than 10^4 ohms per square. (Of course, the shield layer of a shielded container must not normally touch the component or assembly.)

10. Foil backed tapes are *not* to be used for component protection. (Use shielded containers if necessary; see 8 and 9).
11. *All* parts, other than "especially ESD sensitive" parts, should at least be transported, shipped, and stored in containers and packing that meet rules 5 and 6. *This includes plastic parts as well as electronic components*! If this is not possible for some parts, then they must be completely discharged prior to possible contact with other parts.
12. Measure the resistivity of containers, work surfaces, grounds, etc., at periodic intervals to ensure they are still within specifications.
13. "Hunt" for static at regular intervals to verify that the protection system is working.

9 ESD Simulator Design and Usage

Testing product designs to determine their electrostatic discharge (ESD) response is necessary because customers want proof that the design has been properly done, and because designers are not perfect. *The purpose of ESD testing is to ensure equipment will not experience problems during normal usage*. The purpose of the test should not be "to verify we pass the spec." This is an important point because spec and standards authors are not always primarily concerned with the development of a *realistic* test. The following discussion points out those ESD simulator features necessary to simulate a real-world ESD event. These requirements sometimes coincide with existing standards, but often go beyond such standards. As you read this, remember why you actually do ESD tests.

It would be extremely cumbersome and unpleasant to have to perform ESD tests by relying on an actual human being to discharge into the equipment under test (EUT). Therefore, devices have been developed to generate ESD in a manner which simulates real-world conditions. Typically, this means simulation of a human ESD event, but it is also possible to simulate ESD from other items, such as chairs or equipment carts. The following discussion deals with simulators that are intended to simulate a human holding a metal object, which is considered by many to be a good worst-case test.

Because the voltage of the ESD from a simulator is easier to control than for a human, a simulator arc is somewhat more repeatable. In fact, for voltages low enough to prevent significant corona discharge, and for high voltages which have heavy corona effects, arc discharges made with a given tip geometry, speed of approach, and voltage have ESD current waves that are very repeatable. However, even a simulator will not always generate identical arc discharge waveforms, especially at intermediate voltages where corona effects are extremely variable. Therefore, there will still be inherent test variability. In an effort to reduce the variability due to different arcs, some ESD simulators inject charge by direct contact to the EUT. With these direct contact simulators, arc variability can virtually be eliminated and thus ESD testing can be more repeatable. However, even when using contact simulators, some variability in ESD responses will exist. This is because the arc variation is not the only variation that exists from test to test. Circuit signals may be at different levels and states, processors may be executing different instructions, and so on. Unless an EUT can be placed in a perfectly identical state for every discharge, ESD test responses, even for a contact simulator, will appear to be quite variable, and will require statistical analysis.

Although contact ESD simulators may reduce test variability, they do have disadvantages which are related to the inherent requirement that the contact ESD simulator must be conductively connected to the EUT. Many, if not most, products being designed today utilize plastic enclosures. Although these plastic enclosures usually have many openings and seams which could allow a true ESD arc to enter, the openings are often not sufficient to allow insertion of a contact ESD simulator. This makes it difficult to perform a test to determine how the EUT will respond to real-world situations. With an air discharge simulator, one may simply move the simulator over the surface of the EUT to determine where arcs will occur. With a contact simulator, a visual analysis is the only means of determining where discharges *may* occur. After the discharge points are determined, a contact ESD simulator may require extra effort in order to perform the ESD test.

In some cases, removal of the plastic enclosure is all that is necessary to enable contact to be made. However, in many cases the plastic enclosure is required to hold internal assemblies in their proper positions. In these cases technicians may use methods such as soldering wires to internal points and leading the wires out through the enclosure openings or seams. Of course, the wires then add inductance which may slow the rise times of the discharge current waves, and thus reduce the severity of the test. However, the primary concern related to a contact ESD simulator is that there is no air discharge. Without the actual arc in air, the fields generated by a contact simulator are not necessarily identical to true ESD fields. In addition, the contact simulator may generate different electrostatic fields prior to the arc. In the case of real world ESD, the electrostatic field itself can occasionally cause problems. Because the tip of a contact simulator is directly in contact with the EUT, there is no electrostatic field, and thus this damage mode is never tested.

Because of these pros and cons, there are presently two basic approaches to ESD testing.

- Use a contact simulator, which may be more repeatable, but less realistic.
- Use an air discharge simulator, which may have the opposite problems.

A conservative test engineer may wish to use both.

If an air discharge simulator is chosen, the tip of the simulator typically can be either round or pointed. A round tip will have little corona. Thus, it will hold a full voltage until the instant of discharge. A sharp tip will form more corona and allow some voltage to bleed off prior to the actual discharge. Therefore, at all but very low voltages (below a few kV), a rounder tip will usually cause the fastest rise times in the ESD waveform, and will thus provide a worst-case test. A sharp tip should only be used to simulate insertion into EUT openings of sharp tools, keys, ballpoint pens, etc., not for

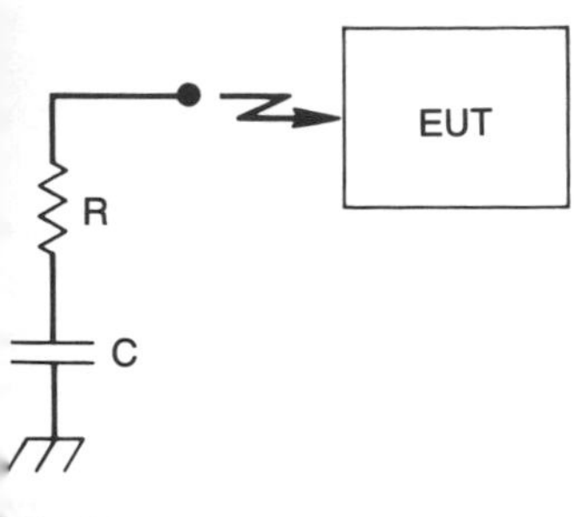

Fig. 36: *Simple RC model of ESD.*

normal worst-case testing of easily accessible areas. For contact simulators, a sharp tip must be used to break through oxide or paint on the metal surface.

Whichever type of simulator is chosen, it is necessary that it provide an output which simulates the primary characteristics of real world ESD.

For many years, ESD simulator designers have modeled the human body as an *RC* circuit (see Figure 36). They therefore designed simulators to look like this model. Eventually, it was realized that small stray capacitance and inductances that had been ignored "for simplicity" were vitally important. For example, the ESD model in Figure 36 theoretically results in an infinite signal rise time, but real ESD, of course, has a finite rise time due to inductance. Therefore, inductance of the ESD simulator wires has to be controlled, and set equal to the human body inductances. Even more important, it must be realized that the human body *cannot* be modeled by a signal *RC*, or even *RLC*, circuit.

As discussed in Chapter One, the circuit shown in Figure 37 is the minimum that can be used to model a human, with reasonable accuracy. This circuit accounts for the main body, the arm, and the hand and finger. Each of these body sections has an important impact on the ESD event. The reader is referred to Chapter One for details.

All ESD simulators have the main body components C_H, R_H, and L_H. The inductance L_H may not be consciously designed in and may just be wire inductance. Where many ESD simulator designs fail is in neglecting to consider the arm, hand, and finger components C_F, C_A, C_{AK}, L_A, and R_A. This prevents these ESD simulators from generating a sufficient initial current spike, which is responsible for a large share of the ESD-related problems.

I. An ESD Simulator Must Be Designed to Simulate the Human Body Inductance (L_H) Properly

II. An ESD Simulator Must Be Designed to Simulate the Human Arm, Hand, and Finger Capacitance, Inductance, and Resistance (C_F, C_{AK}, C_A, L_A, and R_A) Properly

The simulation of C_{AK} is especially difficult because it is referenced to the EUT, and the simulation of C_A is difficult because it is referenced directly to earth. For example, the ESD simulator design in Figure 38 has C_A

Fig. 37: *Electrical model of human ESD.*

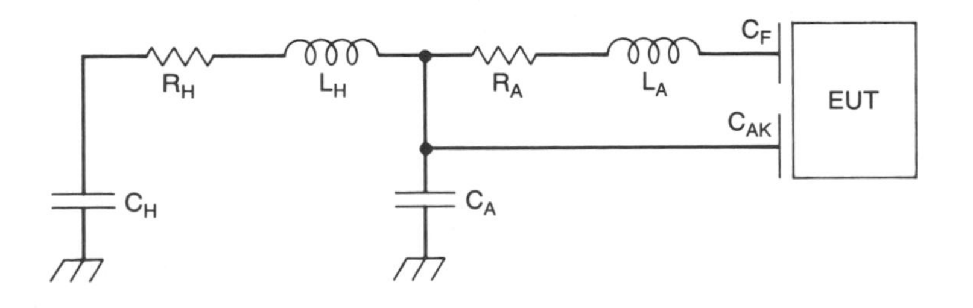

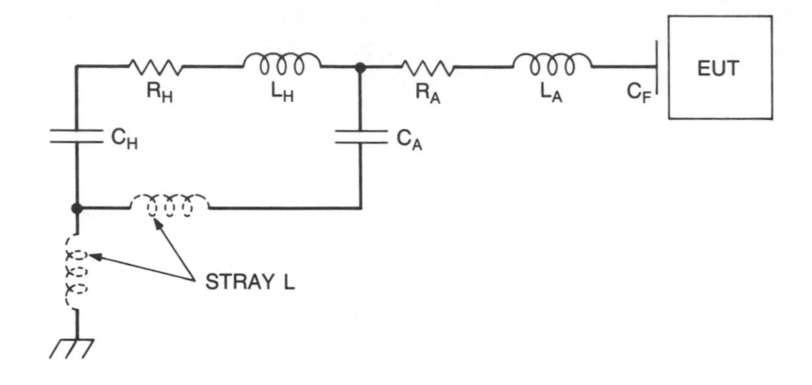

Fig. 38: *Faulty ESD simulator design.*

improperly connected to earth ground *by a wire*, and C_{AK} is not simulated at all.

The ground path for C_A shown in Figure 38 has significant stray inductance! This stray inductance will add to L_A and substantially slow down the discharge of C_A. (In fact, the stray inductance in the ground connection also adds to L_H; this must be accounted for when establishing the value for L_H.) For the simulator to work properly, C_{AK} and C_A must be designed in as a "free space" capacitor; that is, as a sphere, disk, cylinder, etc., that "looks" capacitively like a hand and arm. Just as the hand and arm act as one-half of capacitor C_{AK}, and the EUT acts as the other half, so will a free space capacitor act as one-half of C_{AK}, which is coupled to the other half in the form of the EUT. Also, a free space capacitor will be referred to ground without the need for a wire connection, just as is C_A.

So far, nothing has been said about the exact sizes of R, L, and C. The values of R and L that a person may have vary widely, and the value of C is extremely variable, depending on the proximity to ground (e.g., in the interior of a car, a person's capacitance is very high because he/she is very close to the grounded portions of the car seat). Many electronics standards and specifications give values for R_H and C_H, but none directly specify L_H, C_A, R_A, L_A, and C_F. (C_F is partially specified, however, in an indirect manner by tip dimensions.) As was previously mentioned, at the present time, many ESD simulator producers are using the model of a human holding a metal object (such as a key) as their basis. It is felt that the ESD events which cause the most frequent problems in the electronics industry are related to human contact, and the addition of the metal discharge object, which acts to reduce the skin resistance, and thus R_A, is felt to result in a realistic worst-case model.

The C_H and R_H values presently in standards are listed in Table II for reference (some of these are likely to be revised). Although not listed in any spec, there seems to be some growing agreement on $C_H = 150$ pF and $R_H = 500\ \Omega$ as a standard which realistically simulates a human. (Remember, C_H and R_H are not sufficient by themselves.)

TABLE II

	C_H (pF)	R_H (Ω)
EIA	60	10,000
EIA	100	500
NEMA	100	1,500
MIL	100	1,500
IEC (Present)	150	150
IEC (Pending)	150	330
ECMA	150	330 or 15
SAE	300	5,000

III. Choose the Proper *R*, *L*, and *C* Values for Your Case

Because of the variability of R, L, and especially C in the real world, you should determine if your simulator should use special values. If not, stick with something in the middle, but be sure it represents reality.

Even though the present standards don't specify L_H, C_A, C_{AK}, R_A, and L_A, you still need them to simulate the true ESD event. Values of L_H can range from 0.4 to 2 μH. Values of R_A are in the range of 20 to 200 Ω. Values of L_A can be 0.05 to 0.2 μH. The values of C_A and C_{AK} are, of course, quite variable depending on the location of the free space capacitance. However, the combined free space capacitance of C_A and C_{AK} is on the order of 3 to 25 pF. Of course, the ultimate proof that the values chosen are proper is that the ESD simulator output duplicates the true ESD.

10 Guidelines for ESD Facilities and Test Methods

The selection of the proper electrostatic discharge (ESD) simulator will not alone insure that the ESD test is meaningful. In order to achieve the goal of a realistic worst-case test, a well-defined ESD "test bed" is required.

Any typical lab environment will be an acceptable location for an ESD test area. People are often very concerned about controlling environmental factors such as temperature, barometric pressure, and humidity. Although humidity has a large impact on the development of true static charge, humidity has little effect on the charging of an ESD simulator. The ESD simulator, unlike a person, has a high voltage power supply to ensure it is fully charged. The primary impact of humidity is to encourage or discourage corona, and thus impact the rise time of the discharge current. However, the humidity control in a typical lab should be sufficient to prevent significant variations in corona and thus rise time. Temperature and pressure are of low importance in the development of true ESD, and they are even less important to ESD simulator test results.

The primary problem that may occur when an ESD test area is placed in a normal lab is the effect of ESD testing on other lab activities. Steps must be taken to isolate the ESD effects to the test area.

I. Keep the Boundary of the ESD Test Area At Least One Meter Away from Other Test Equipment and Wiring (This Includes the AC Power Wiring in the Walls, Ceiling, and Floors)

II. An Isolation Transformer Should Be Used to Supply AC Power to the ESD Simulator and Equipment Under Test (EUT)

Note: To meet electrical code requirements, the safety "green" ground must not be disconnected from the simulator or the EUT.

A. The isolation transformer must be as close as possible to the AC outlet feeding it.
B. The power cord from the isolation transformer to the ESD simulator and EUT must rest on an earth ground plane (see next guideline), and should be as short as possible.

III. There Must Be a Large Ground Plane Covering the Floor of the ESD Test Area

This floor ground plane must be connected by a short connection to the safety (green) ground of the AC wiring and the isolation transformer. Also, the shield of the isolation transformer should be connected to the ground plane by a short connection.

These connections are required for safety, and to allow the plane to "filter" ESD noise from the safety ground. This large ground plane acts as a large "free space" capacitor (approximately 30 pF per square foot) that can absorb ESD charge, and then slowly bleed it off into the green ground of the building's AC wiring. In addition to filtering the ESD currents that enter the green ground, this floor ground plane acts as a shield for any wiring in the floor under it.

Just as the floor ground plane helps isolate the AC wiring from the effects of ESD, it also isolates the ESD test bed from unknowns in the earth ground system of the building. For example, the total impedance of the earth ground path is seldom known and depends on the conductivity of earth, which may vary. Also, changes in moisture levels can change the conductivity of concrete floors, which may change the stray capacitance from earth to the EUT. The floor ground plane ensures that the stray capacitances that exist in the ESD test area will remain constant.

Although the isolation effect of the ground plane is clear, what is less clear is whether the use of a ground plane will result in a more, or less, severe test. In order to understand the impact of a ground plane, refer to the ESD model in Figure 39. (This was developed in Chapter 1.)

In this case C_H, L_H, R_H, C_{AK}, R_A, L_A, and C_F are all determined by the design of the ESD simulator. Items L_S and R_S are determined by the arc itself, and items R_K and L_K are determined by the ground path of the EUT. (In previous examples the EUT was a keyboard.) The primary effect of the

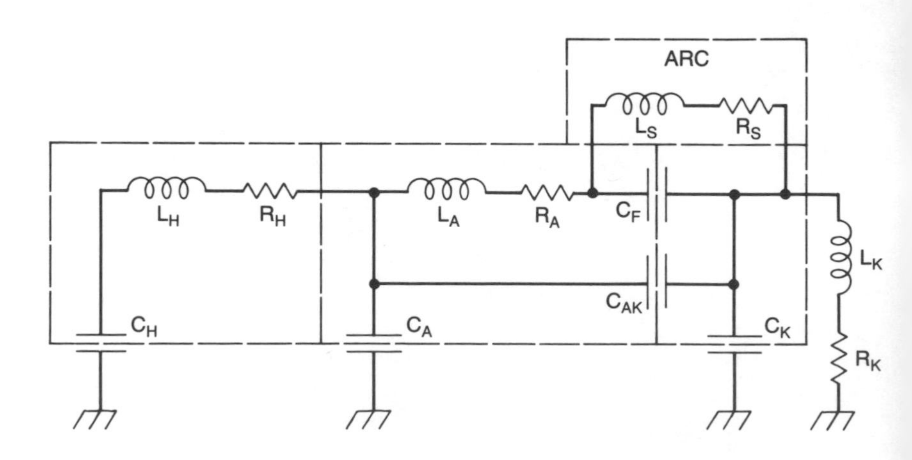

Fig. 39: *Human and EUT model for ESD.*

ground plane is to increase the value of C_A and of C_K which effectively shunts around the normal chassis ground path (R_K and L_K). An increase in the value of C_K will result in less attenuation for the higher frequency components of the ESD current. (For a more complete discussion of the ESD model and ESD effects, see Chapters 1 and 2, respectively.) As a result, the high frequency *H*-fields will be increased and *E*-fields will be reduced. The impact of this on the test results will depend on whether the EUT is more sensitive to *H* or *E* fields. Experience indicates that the majority of systems will see this as a more severe test.

This increase in severity will be reduced somewhat by the reduction of high frequency ESD components in the chassis ground cable. However, most cable noise problems are associated with lower ESD frequencies, so the reduction of high frequency cable noise will typically be insufficient to counter the harmful effects of high frequencies elsewhere.

From an ESD standpoint, the material from which the ground plane is constructed is not very important. The ground plane thickness should be at least three "skin depths" over the entire ESD frequency range, but this is easily provided. The only critical requirements have to do with corrosion and strength.

A. Bonds to ground planes should use materials with less that 0.25 V EMF difference in the electrochemical series.
B. The ground plane must support equipment for ESD testing without ground plane deformation or damage.
C. For safety purposes, the floor ground plane should be covered with a high resistance walking surface to limit fault current to personnel to less than 0.5 mA.
D. The floor ground plane must be at least 2.5 m × 2.5 m.

If the EUT is to be used on a desk or tabletop, it is convenient to test it on a table (and it is also more realistic). In this case, a nonconductive table is placed on the floor ground plane. (See the example of an ESD test facility in the appendix, i.e., Figure 40, page 94.) The table structure should be nonconductive, so it doesn't create additional reflections and/or coupling paths. Since work surfaces in the real world can have metal or metal-reinforced surfaces which are grounded, the ESD test table also needs a metal ground plane which can be installed to simulate metal tabletop situations.

IV. Just As With the Floor Ground Plane, a Table Ground Plane Option is Required in Order to Achieve a Realistic, Repeatable, and Worst-Case Test

A. The tabletop ground plane bonds must use materials with less than 0.25 V EMF difference in the electrochemical series.
B. The tabletop ground plane must be at least 10 cm more in length and

width than the EUT. This ensures that the stray capacitance is not limited by the table area.

C. The tabletop ground plane must support the EUT without deformation or damage.

D. Since most work surfaces are not made of metal but rather are reinforced by metal, the tabletop ground ground plane may be covered with a nonconductive work surface (see note below). However, this should only be done if it is certain that the EUT will never be placed on a metal tabletop. Some ESD specs call for the EUT to be placed on blocks, so air is the dielectric between the EUT and table ground plane. A full nonconductive surface is better because it has a higher dielectric constant than air. This higher dielectric constant will increase the stray capacitance from the EUT to the table ground plane. The thickness of the nonconductive surface can also affect the ESD test results. A thin dielectric will increase the stray capacitance of the EUT to the ground plane, and thus generally result in a harsher test. At present, a 2.5 cm thickness is often specified for the nonconductive surface; however, there is some movement toward thinner surfaces. One must be careful, however, if a thin plastic dielectric is chosen, because it may develop pin holes with use.

   Note: Nonconductive surfaces other than wood may be advisable if the humidity varies significantly, because the wood may change its conductivity with changes in humidity.

Before discussing the manner in which the tabletop plane is connected to the earth ground, the concept of ''indirect'' ESD testing needs to be explained. Up to this point, the description of ESD testing has been for direct tests; that is, the ESD simulator discharges directly to the EUT. In the real world, ESD can also occur indirectly. For example, a person can discharge to a nearby metal object such as a desk or file cabinet near a desktop computer. In this case there is no direct charge injection to the EUT itself, but the EUT will be impacted by the fields radiated from the nearby metal object. Thus, even if a particular EUT is well insulated so that no arc to the EUT can occur, an indirect test is still necessary to verify that the EUT is ESD immune.

To perform an indirect test, the ESD simulator is discharged to a metal object which is near the EUT. For safety, and to ensure that each trial is independent, this metal object must have a path to bleed injected charge to ground. In the real world, however, earth ground paths are nearly always relatively high impedance at ESD frequencies (often real desks, file cabinets, and tables aren't grounded at all). Therefore, the metal object that is used as a target for the indirect ESD test should normally have a ground connection which will simulate the nongrounded object. This will most closely simulate real-world conditions by encouraging the generation of standing waves on the metal object. The standing waves are created because of the impedance mismatch between the metal object and ground path. The larger these standing waves, the more intense the fields radiated by the metal object.

This metal object can be oriented horizontally or vertically and is typically a flat surface. When doing the indirect test with a horizontal metal surface, the tabletop ground plane usually serves as the target for the indirect discharge. This seems to result in a conflict about how the tabletop ground plane should be connected to earth ground. In order to simulate the real world, for indirect tests, the impedance of the ground connection should be high; but in order to act as a good ground plane, during direct tests, the impedance from the table to earth should be low. In fact, the problem is more apparent than real. Because of its area, the table ground plane has a significant free space capacitance, and just like the floor ground plane, it will act as a "local" ground in spite of a high impedance connection to earth.

The exact R, L, and C values used to achieve the proper impedance level for the ground connection must remain constant from test to test. Therefore, it is common to use the local earth ground of the floor plane as a connection point for the table plane. Also, it is a good idea for the same values to be used in different ESD test beds. This will result in more uniform tests from facility to facility.

At the present time, there are two preferred values for the resistance of the tabletop to earth ground path. These values are 10 kΩ and 1 MΩ. The capacitance of the connection to earth ground from the table is usually either 100 pF or no capacitance is recommended. The inductance value of the ground path is virtually never specified, except that a very large or wide conductor is often called for to keep inductance very small.

The following guideline meets the two most common sets of requirements.

E. *Option one:* The table ground plane is connected to the floor ground plane by a 15 cm wide conductor. The floor termination then uses a connection which has a capacitance of 100 pF and a resistance of 10 kΩ. The capacitance and resistance should be evenly distributed across the width of the conductor. The length of the conductor should be the minimum necessary to reach from the tabletop plane to the floor plane. *Option two:* The table ground plane is connected to the floor ground plane via a heavy gauge wire. A 1 MΩ resistor is then used to connect the table plane to the wire.

F. In either case, the ground connection to the tabletop ground plane must be as far as possible from the discharge point on the table plane so that maximum standing waves exist over the maximum area. For any indirect or direct test, the discharge point chosen must result in the maximum possible fields near the EUT. This will be discussed further below.

It was mentioned above that the discharge target in an indirect test can be oriented horizontally or vertically. This will be discussed in more detail, but it should be noted that a vertical metal plate should generally follow the same guidelines as the tabletop ground plane for the same reasons. However, since the vertical plate would be simulating items other than tables or desks, the capacitance of the ground connection will normally be different than 100 pF, to reflect the different stray capacitance of the objects simulated.

Once the basic test bed has been constructed, the ESD simulator must be added. The following guidelines cover the interaction of the simulator and test bed.

V. For Worst-Case Testing, the Discharge Point for Either Direct or Indirect Tests Should Be Chosen to Place the Maximum Redistribution Currents and Fields As Near As Possible to the Most Sensitive Electronics

Typically, this will be accomplished by selecting a discharge point away from the chassis ground connection, but near the electronics. This *does not* mean that the EUT should be disassembled to find the "ideal" discharge point. That would not be a realistic test.

VI. The Simulator Should Be Plugged into the Same Isolated AC Source as the EUT

This is done to increase the common-mode isolation from the rest of the lab. The AC outlet for both the EUT and ESD simulator should be supplied by an isolation transformer. The safety "green" ground must pass through to the simulator and EUT.

VII. The ESD Simulator Chassis Ground Must Be Connected to the *Floor* Ground Plane

In the ESD model, the human body capacitance is referenced to earth ground, and the simulator should also be referenced to earth ground. It used to be common to connect the ESD simulator chassis ground to the EUT; this was done because ESD testers used to have a simple *RC* circuit which didn't simulate the hand and arm capacitances. By connecting the simulator chassis ground to the EUT, the normal EUT ground path was bypassed. Thus, the higher frequency simulator currents were much less attenuated, the *H*-fields were increased, and the test was generally more severe. This approach had limited success, however, and it was eventually realized that the choice of the ground connection could not overcome the fundamental problem. Only the addition of a hand and arm simulation could create that important initial current spike that characterized ESD. As well, by connecting the ESD simulator ground to the EUT, and bypassing connection cables, it was not always possible to find problems associated with the cables. A separate ESD test was required for the cables themselves. But, of course, separate tests for each portion of the EUT system cannot necessarily find problems that result from a combination of factors. Therefore, the ESD simulator chassis must not be connected to the EUT, or to the table ground plane, but to the floor ground plane.

VIII. Portions of the ESD Simulator Other than the Discharge Tip Itself (such as Control Boxes) Must Be Kept At Least 0.1 Meter from the EUT

This requirement is not extremely critical, but test repeatability is generally improved by keeping other potential sources of noise, or noise reflection, away from the EUT. For tabletop tests, the best place to locate the simulator controller, etc., is under the tabletop ground plane. The table plane will then isolate the EUT and simulator more completely.

Once the ESD simulator is set up, and a discharge point is selected, the discharge itself should be done properly.

IX. Ensure that the ESD Simulator is Charged to the Proper Level for Each Discharge

A. For air discharge, don't hold the simulator tip so near to the EUT, or other metal, that corona prevents full recharge.
B. Don't discharge so frequently that full recharge isn't possible.

X. Position the Simulator in a Consistent Position Relative to the EUT or Metal Plane During Approach and Discharge. For Air Discharge Simulators, Approach Speed Must Be Consistent from Discharge to Discharge

Usually, in simulators which incorporate a probe, the probe should be perpendicular to the EUT. The position of the ESD simulator determines the value of the free space capacitor that simulates the hand and arm. This capacitance value is important for repeatable tests, and thus the position of the simulator must be uniform from test to test. Also, any cables associated with an ESD simulator can radiate (even though they are usually shielded); therefore, the cable positions should be consistent from test to test.

Although a fast approach will result in the fastest current rise times and peak current levels for an air discharge simulator, it is more difficult to maintain consistent fast approach speeds. The approach speed used should be relatively rapid, but realistic and consistent. In order to hold the air discharge simulator in a stable position and have a consistent approach speed, it may be necessary to use a fixture. If this is done, the fixture should be nonconductive so it doesn't itself add another variable to control. An alternate approach is to hold the ESD simulator position steady and allow the tip to charge until it arcs. In this case the rate of recharge corresponds to the approach speed. This method is fine as long as the EUT can be checked for proper operation between each discharge. Otherwise, the tests are not truly independent (see

Chapter 11). Also, the rate of simulator discharge should be slow enough to allow charge to bleed off the EUT between tests. (A half or quarter second should be sufficient.)

The main problem with the alternate method is that the voltage is adjusted by setting the proper gap width. At lower voltages, the gap width setting must be very precise.

The position of the EUT is just as important as the simulator position.

XI. The Stray Capacitance of the EUT to Ground Must Be Uniform from Test to Test

A. The ground planes should be at least 30 percent larger in area than the EUT.
B. The coupling between the EUT and ground plane should be controlled by using the same dielectric spacer for all tests. The EUT must not extend beyond the dielectric spacer.

XII. When Performing Indirect Tests, the EUT Must Be Positioned so the Maximum Redistribution Currents Flow Past, or Under, the EUT

A. Generally, the plane receiving the indirect discharge should be parallel to the majority of the PWBs within the EUT.
B. Interconnect cables for EUT components should be coiled in the largest possible loop and oriented so the coil plane is parallel to the indirect target's plane.

The EUT should not only be positioned properly, but should be physically configured properly.

XIII. The EUT Must Be Physically Configured to Simulate Realistic Conditions

A. I/O devices, and especially cables, must be installed or must have cables with dummy loads.
B. Shielding and other ESD protection features must be in place.
C. The equipment should be in the operational mode which reflects real operation.
D. Partial or subassemblies should be tested in conditions which simulate the final assembly.
E. If the EUT is to be frequently handled or used while not connected to any other unit or AC power, then it should also be tested in that configuration. If the unit is only transported while disconnected, and not actually operated, then only a destruction rest is required with the unit disconnected.

The remaining ESD test guidelines relate not to hardware, but to *procedures*. Without proper procedures, the best ESD test bed in the world is nearly worthless.

XIV. Clearly Define the Points to Which Discharge is to Occur

In defining test points consider what areas of the EUT are accessible to operators or other personnel. If it is desired that a worst-case test point be selected, this worst-case point should be based on a statistically meaningful sample size to ensure the test is a worst-case test.

As was previously mentioned, the discharge point should subject the sensitive components to the maximum currents and fields. Most present specifications recommend that a pretest be done to determine which discharge point will cause the most problems. This sounds good in theory, but it's not usually done properly in practice. A well-designed piece of equipment will have a low ESD failure rate. Therefore, in order to determine the ''weak spot'' with any statistical reliability, a large number of discharges are typically required. Since the pretests typically used by most people don't use a sample size sufficient to be accurate, the discharge point will essentially be chosen at random. Pretests must be done properly if they are to be meaningful. (Determination of sample sizes is discussed further in Chapter 11.)

XV. Use Unambiguous Failure Definitions That Represent True Real-World Failures

Don't spend good money to eliminate nonproblems, such as temporary LED indicator flicker. Don't use ambiguous terms like ''ESD shall not cause errors'' (see also Chapter 11).

XVI. Use a Statistically Derived Sample Size and Acceptance Limit

ESD results are not black and white. The chances that the electronic circuits will be in exactly the same state for every discharge are very low. A discharge may occur exactly at the sample time for an input, or not. The microprocessor may be executing the same instruction and have the same memory contents, or not. Signal lines may be at the same level and state, or not. ESD testing is a probabilistic process. Given enough discharges, virtually any system may eventually experience a failure of some type. You have to establish how far you want to go (see Chapter 11), and perfection is not a reasonable goal.

XVII. The Test Should Proceed from Indirect to Direct, and from Low Voltage to High Voltage

In order for samples to be independent and accurate, the EUT must not be degraded. Performing the least damaging tests first will help prevent inaccurate test results caused by component degradation (see also Chapter 11).

Both direct and indirect tests are usually necessary because the fields will vary in each case. Both low and high voltages are required because low voltages, which have less corona, have higher frequency components which generate more intense fields, and high voltages have more energy and thus more potential to cause physical damage.

XVIII. Make Sure Any Errors Will Be Detected and that the EUT is Returned to Full and Original Operation After Each Discharge

In order to speed the test process, many technicians will hit the EUT with a very fast burst of discharges (possibly even thousands per burst). There are two problems with this approach. First, the charge may not have time to bleed off the EUT between discharges; this means the voltage level of the EUT rises, and subsequent discharges in a burst are effectively at a lower voltage than initial discharges. Second, and much more important, accurate error detection during rapid bursts is extremely difficult, if not impossible. For example, an initial discharge may reset the program counter of a microprocessor, and a subsequent discharge could erase a bit of memory. Unless the error detection method is much faster than the discharge burst rate, these multiple errors will be counted as one error. Such miscounts may occur even for very serious errors. For example, one discharge may cause complete IC lockup, but the next may reset the IC; if this happens rapidly, no error may be detected.

The above examples of the inaccuracies in error detection will result in an undercount of the errors. However, it is also possible to overcount. Unless the error detection system is able to find and clear the cause of the errors detected, they may be counted repeatedly. For example, if a register is set to an incorrect value by ESD, an error will occur every time the register value is used.

Ensuring the accuracy of error counts relates to the problem of recording the error count in a secure manner. Ironically, automatic error detection and recording systems often rely on the very system that is being tested, in order to detect and keep track of its own errors. If this is done, it must be certain that the ESD test won't cause errors in the detection count/storage system itself.

In order for test results to be accurate, there must be sufficient time between discharges to allow the errors to be detected and securely recorded, and to completely restore the EUT to the original operating condition.

XIX. Stop the Test as Soon as the Acceptance Limit is Exceeded

This saves time as well as wear and tear on the equipment. (However, it must be remembered that ESD testing may itself cause latent failures, especially if it is direct testing. Therefore, if the EUT is to be shipped, it may be wise to limit ESD test stresses.)

XX. Periodically Check the "Calibration" of the ESD Test Bed

Because of the very high frequencies and short-lived signals involved, directly measuring signals isn't at all simple. A quick and dirty method is to have two "standard EUTs" that have a known failure rate. These standard EUTs can be tested every few months to determine if the failure rate shows a statistically significant variation. If it does, and the changed results of the standard EUTs agree with each other, the test bed ground connections, etc., must be checked. If facility grounding, etc., isn't a problem, then your simulator probably has changed. In this case, you will probably have to be aided by the firm that produces the simulator. Fortunately, ESD simulators are not typically prone to failure, and the calibration check will just be a verification that all is well. Needless to say, the standard EUTs should never be tested directly or at destructive levels. This will help ensure that the standard EUTs aren't damaged and don't change their ESD response.

XXI. Be Careful

ESD tests involve extremely high voltages, and, even though they aren't necessarily life threatening, discharges at 25 kV are certainly not good for the human system. Also, because of the fields involved, people with electronic life support items, such as pacemakers, probably shouldn't be near ESD testing.

11 Statistical Sampling Criteria

Over the past several years electrostatic discharge (ESD) specifications have become more and more detailed, in an attempt to obtain accurate and repeatable results. Some years ago it was not uncommon for a specification to consist of a single sentence such as, "The unit shall exhibit no failures due to ESD at a level of 5 kV." Presently, specifications often specify the construction of the simulator, the physical test set-up, and even the environmental conditions, such as humidity and temperature. Instead of a single ESD level, present specifications often list various failure rates at various voltage levels.

Why Most Tests Don't Work

While it is true that the physical test set-up and simulator design are very important factors, it is unfortunate that most people have focused on these factors, while ignoring the equally important statistical requirements of the sampling process. This lack of knowledge results in the failure of the test, no matter what physical set-up or simulator is used, and thus ESD testing is often seen as "black magic"! The reason for this continuing problem is simple. *Although nearly everyone recognizes the random occurrence of ESD problems, most ESD tests are not constructed by applying the appropriate statistical concepts.* The determination of proper sample sizes and the proper statistical analysis of test results are two of the least understood aspects of ESD testing. This is unfortunate because the use of proper statistical methods is not just an esoteric exercise, but a real necessity, in order to achieve reliable test results. Table III is a typical (but, unfortunately, unacceptable) example of an ESD specification.

The primary problem with this example is that the author didn't formally establish a desired confidence level for his test results. If he had, he would have realized that 50 trials is too small to have an acceptable confidence level at failure rates such as 2 percent.

Also, had the author attempted to calculate a confidence level, he would have realized that a measured failure rate of zero doesn't mean the true failure rate is zero. In fact, verification that the true failure rate is zero percent would require an infinite test. It is no surprise that a unit could pass this "typical" test at times and fail it at other times. In fact, one would expect that this would happen. When such tests are used to make decisions about which product to

TABLE III: *Typical, but unacceptable, ESD specification*

Data Errors		Lock-Ups	Damage	No. of Discharges
5 kV	0%	0%	0%	50
10 kV	2%	0%	0%	50
15 kV	20%	2%	0%	50
20 kV	50%	10%	0%	50

purchase, or to determine future design guidelines, it is clear that the decisions will often be faulty.

How Sure and How Good?

It is not possible to be 100 percent confident of test results. The question is, how confident do you *need* to be? Remember, if you are 90 percent confident, there's still a one in ten chance that you're wrong. How much risk are you willing to accept? Likewise, *it is not possible to verify that the true failure rate is zero percent*. Again, the question should be, how low does the true failure rate really *need* to be?

These questions must usually be answered based on market inputs tempered by the overall quality philosophy of the company. It is obvious that high confidence and low failure rates are desirable to the customer, but these desires also result in a requirement for more trials. Extremely large numbers of trials could make the test impractical.

Once the confidence levels and maximum failure rates are established, then, and only then, can the test parameters be established. In general, ESD tests are usually done for two reasons: (1) to verify the unit does, or doesn't, exceed acceptable limits (acceptance testing); and (2) to compare the response of two or more units (comparative testing). Before discussing the requirements for each of these tests, one must understand confidence levels and confidence limits.

Confidence Levels and Limits in Brief

Properly done ESD test results conform closely to a binomial distribution. The criteria for such a distribution are:

1. There are only two possible outcomes for each trial (success and failure).
2. The probabilities of success and failure are constant from trial to trial.
3. There are N trials.
4. The N trials are independent.

The above criteria imply that failure and success must be defined properly so that only one or the other is possible. Also, different conditions must not be allowed during different trials, so the EUT must be discharged to its base condition between trials, and full EUT operation must always be restored after any failure. For practical purposes, damage may be allowed as long as the EUT is completely repaired before the next trial. Of course replacement parts must be identical (e.g., vendor and vendor part number) to the original parts.

Proper definition of success and failure is important. Success is typically defined as the absence of failure; however, this leaves the problem of defining failure. The key is to be complete and precise in indicating the response of the unit to ESD. If the failure definition is nebulous, then confidence in the numerical test results is immaterial. You still haven't characterized the ESD response of the unit. In most cases, multiple failure categories are desirable. In order not to violate the binomial criteria, each type of failure must be considered independently. From the standpoint of each failure type, each test trial is considered to have only two outcomes: Either that particular failure happens, or it doesn't. This results in some apparently odd situations. For example, when testing for "soft" failures, destructive failures could be considered "successes." It all depends on your failure definition.

The binomial distribution becomes more difficult to work with as the number of trials increases. Therefore, it is common to approximate the binomial distribution with either a Poisson or a Normal distribution. For less than 10 failures and more than 100 trials, the Poisson is better; for more than 20 trials, and more than a 5 percent failure rate, the Normal distribution is better. When the failure rate of an EUT is totally unknown, the basic confidence limit formulas are as follows: (As you read this, keep in mind that there is a given confidence *level*, and that the true failure rate is within the calculated confidence *limits*.)

Poisson

$$P = \frac{X^2}{(2)(N)}$$

P = The confidence *limit* for the failure rate. (This corresponds to the maximum true failure rate.)

N = The number of trials. The value of the term X^2 is found from Table IV. In Table IV, the desired confidence *level* is listed across the top (a confidence *level* of 1.0 equals 100 percent confident). The left column (V) is found by the formula $V = 2(F + 1)$, where F is the actual number of failures encountered during the test. This formula means that, if there are F failures out of N trials, then one may be some given percent confident (confidence *level*) that the true failure rate is less than P (confidence *limit*). This formula, or variations on it, is the basis for calculations when the failure rates are low.

TABLE IV: *Values of X^2 Confidence Level*

V	0.005	0.01	0.025	0.05	0.95	0.975	0.99	0.995
1	0.0000393	0.000157	0.000982	0.00393	3.841	5.024	6.635	7.879
2	0.0100	0.0201	0.0506	0.103	5.991	7.378	9.210	10.597
3	0.0717	0.115	0.216	0.352	7.815	9.348	11.345	12.838
4	0.207	0.297	0.484	0.711	9.488	11.143	13.277	14.860
5	0.412	0.554	0.831	1.145	11.070	12.832	15.086	16.750
6	0.676	0.872	1.237	1.635	12.592	14.449	16.812	18.548
7	0.989	1.239	1.690	2.167	14.067	16.013	18.475	20.278
8	1.344	1.646	2.180	2.733	15.507	17.535	20.090	21.955
9	1.735	2.088	2.700	3.325	16.919	19.023	21.666	23.589
10	2.156	2.558	3.247	3.940	18.307	20.483	23.209	25.188
11	2.603	3.053	3.816	4.575	19.675	21.920	24.725	26.757
12	3.074	3.571	4.404	5.226	21.026	23.337	26.217	28.300
13	3.565	4.107	5.009	5.892	22.362	24.736	27.688	29.819
14	4.075	4.660	5.629	6.571	23.685	26.119	29.141	31.319
15	4.601	5.229	6.262	7.261	24.996	27.488	30.578	32.801
16	5.142	5.812	6.908	7.962	26.296	28.845	32.000	34.267
17	5.697	6.408	7.564	8.672	27.587	30.191	33.409	35.718
18	6.265	7.015	8.231	9.390	28.869	31.526	34.805	37.156
19	6.844	7.633	8.907	10.117	30.144	32.852	36.191	38.582
20	7.434	8.260	9.591	10.851	31.410	34.170	37.566	39.997
21	8.034	8.897	10.283	11.591	32.671	35.479	38.932	41.401
22	8.643	9.542	10.982	12.338	33.924	36.781	40.289	42.796
23	9.260	10.196	11.689	13.091	35.172	38.076	41.638	44.181
24	9.886	10.856	12.401	13.848	36.415	39.364	42.980	45.558
25	10.520	11.524	13.120	14.611	37.652	40.646	44.314	46.928
26	11.160	12.198	13.844	15.379	38.885	41.923	45.642	48.290
27	11.808	12.879	14.573	16.151	40.113	43.194	46.963	49.645
28	12.461	13.565	15.308	16.928	41.337	44.461	48.278	50.993
29	13.121	14.256	16.047	17.708	42.557	45.722	49.588	52.336
30	13.787	14.953	16.791	18.493	43.773	46.979	50.892	53.672

Normal

In the case of the normal distribution, the confidence limit is

$$P = F/N + Z\sqrt{\frac{F/N(1-F/N)}{N}}.$$

As before, P = maximum failure rate limit (confidence limit); N = number of trials; F = the actual number of failures; and Z comes from Table V, based on the confidence level. To find Z, locate the confidence level in the table, and then refer to the associated left column and top row to get the value for Z (e.g., for 97.5 percent confidence, Z = 1.96). This confidence limit formula, or variation of it, is the basis for statistical calculations when the failure rate is higher.

TABLE V: *Values of Z Confidence Level*

↓Z	0.00	0.01	0.02	0.03	0.04	0.05	0.06	0.07	0.08	0.09
0.0	0.5000	0.5040	0.5080	0.5120	0.5160	0.5199	0.5239	0.5279	0.5319	0.5359
0.1	0.5398	0.5438	0.5478	0.5517	0.5557	0.5596	0.5636	0.5675	0.5714	0.5753
0.2	0.5793	0.5832	0.5871	0.5910	0.5948	0.5987	0.6026	0.6064	0.6103	0.6141
0.3	0.6179	0.6217	0.6255	0.6293	0.6331	0.6368	0.6406	0.6443	0.6480	0.6517
0.4	0.6554	0.6591	0.6628	0.6664	0.6700	0.6736	0.6772	0.6808	0.6844	0.6879
0.5	0.6915	0.6950	0.6985	0.7019	0.7054	0.7088	0.7123	0.7157	0.7190	0.7224
0.6	0.7257	0.7291	0.7324	0.7357	0.7389	0.7422	0.7454	0.7486	0.7517	0.7549
0.7	0.7580	0.7611	0.7642	0.7673	0.7704	0.7734	0.7764	0.7794	0.7823	0.7852
0.8	0.7881	0.7910	0.7939	0.7967	0.7995	0.8023	0.8051	0.8078	0.8106	0.8133
0.9	0.8159	0.8186	0.8212	0.8238	0.8264	0.8289	0.8315	0.8340	0.8365	0.8389
1.0	0.8413	0.8438	0.8461	0.8485	0.8508	0.8531	0.8554	0.8577	0.8599	0.8621
1.1	0.8643	0.8665	0.8686	0.8708	0.8729	0.8749	0.8770	0.8790	0.8810	0.8830
1.2	0.8849	0.8869	0.8888	0.8907	0.8925	0.8944	0.8962	0.8980	0.8997	0.9015
1.3	0.9032	0.9049	0.9066	0.9082	0.9099	0.9115	0.9131	0.9147	0.9162	0.9177
1.4	0.9192	0.9207	0.9222	0.9236	0.9251	0.9265	0.9279	0.9292	0.9306	0.9319
1.5	0.9332	0.9345	0.9357	0.9370	0.9382	0.9394	0.9406	0.9418	0.9429	0.9441
1.6	0.9452	0.9463	0.9474	0.9484	0.9495	0.9505	0.9515	0.9525	0.9535	0.9545
1.7	0.9554	0.9564	0.9573	0.9582	0.9591	0.9599	0.9608	0.9616	0.9625	0.9633
1.8	0.9641	0.9649	0.9656	0.9664	0.9671	0.9678	0.9686	0.9693	0.9699	0.9706
1.9	0.9713	0.9719	0.9726	0.9732	0.9738	0.9744	0.9750	0.9756	0.9761	0.9767
2.0	0.9772	0.9778	0.9783	0.9788	0.9793	0.9798	0.9803	0.9808	0.9812	0.9817
2.1	0.9821	0.9826	0.9830	0.9834	0.9838	0.9842	0.9846	0.9850	0.9854	0.9857
2.2	0.9861	0.9864	0.9868	0.9871	0.9875	0.9878	0.9881	0.9884	0.9887	0.9890
2.3	0.9893	0.9896	0.9898	0.9901	0.9904	0.9906	0.9909	0.9911	0.9913	0.9916
2.4	0.9918	0.9920	0.9922	0.9925	0.9927	0.9929	0.9931	0.9932	0.9934	0.9936
2.5	0.9938	0.9940	0.9941	0.9943	0.9945	0.9946	0.9948	0.9949	0.9951	0.9952
2.6	0.9953	0.9955	0.9956	0.9957	0.9959	0.9960	0.9961	0.9962	0.9963	0.9964
2.7	0.9965	0.9966	0.9967	0.9968	0.9969	0.9970	0.9971	0.9972	0.9973	0.9974
2.8	0.9974	0.9975	0.9976	0.9977	0.9977	0.9978	0.9979	0.9979	0.9980	0.9981
2.9	0.9981	0.9982	0.9982	0.9983	0.9984	0.9984	0.9985	0.9985	0.9986	0.9986
3.0	0.9987	0.9987	0.9987	0.9988	0.9988	0.9989	0.9989	0.9989	0.9990	0.9990
3.1	0.9990	0.9991	0.9991	0.9991	0.9992	0.9992	0.9992	0.9992	0.9993	0.9993
3.2	0.9993	0.9993	0.9994	0.9994	0.9994	0.9994	0.9994	0.9995	0.9995	0.9995
3.3	0.9995	0.9995	0.9995	0.9996	0.9996	0.9996	0.9996	0.9996	0.9996	0.9997
3.4	0.9997	0.9997	0.9997	0.9997	0.9997	0.9997	0.9997	0.9997	0.9997	0.9998

Proper Sampling for Units with Low Failure Rates

The definition of acceptance limit needs to be well understood at this point. It was previously stated that the maximum allowed true failure rate was a joint technical/managerial decision. Assume a company decides 2 percent is the maximum true failure rate they will accept. Since there is always some variability involved with tests, it should be clear that the test acceptance limit shouldn't be set at exactly 2 percent. There should be a safety margin to ensure the item tested is really below 2 percent in spite of test variability. For example, if the safety margin is 0.5 percent, then the acceptance limit should be 1.5 percent. This will insure that an item with a true 2.5 percent failure rate will not be passed, just because ±0.5 percent test variability resulted in a 2 percent measured failure rate. In this case, if the maximum acceptable true failure rate is 2 percent, the acceptance limit should be 1.5 percent.

Testing at low failure rates is virtually never done properly, so detailed instructions for *acceptance testing* when the failure rate is low are given below.

1. *The confidence level and the maximum failure rate* (P) *should have been established before the testing begins*!

2. Next establish the number of trials (N), and number of actual failures (F) allowed in N (the acceptance limit). Do this by rearranging the formula for the Poisson confidence limit.

$$N=\frac{X^2}{2(P)}\ .$$

3. The previous formula is based on two unknowns (N and F). In order to find N, we must first establish the number of actual failures to be allowed during the test (F). As mentioned, the value of F is the acceptance limit for the test. The higher the value of F is, the larger N will have to be for a given confidence level. Therefore, it is reasonable to simply set $F = 0$. This will always result in the minimum N.

Example: Desired confidence level = 95%.

Maximum true failure rate accepted, $P = 0.05$.

Set $F = 0$; so, $V = 2(F + 1) = 2$, and from Table IV, for $V = 2$ and confidence = 0.95, $X^2 = 5.99$.

Therefore,

$$N=\frac{5.99}{2(0.05)}=59.9$$

which rounds to $N = 60$.

To test for a 5 percent maximum true failure rate, therefore, would require at least 60 trials, and there could be no failures. That is, the acceptance limit is zero out of 60, if one wishes to be sure that the true failure rate is under 5 percent. Note that this doesn't prove the failure rate for the unit tested is exactly 5 percent, but that it is most likely *less than* 5 percent. To know how much less, retest for a reduced P (and thus increased N).

Proper Sampling for Larger Failure Rates

For larger failure rates, such as 10 percent, 20 percent, 50 percent, etc., the Normal distribution is used in place of the Poisson because the Normal more closely approaches the binomial distribution at these levels. Again, for *acceptance testing* rearrange the confidence limit formula to solve for N.

$$N = P(1-P)\left[\frac{Z}{P-M}\right]^2.$$

Also, as before, this formula is based on two unknowns. This time, however, instead of setting a value for the allowed number of failures (F), set a percentage limit of M for the measured failure *rate*. In this case M is the acceptance limit.

An example follows:
Desired confidence level of 95 percent; desired acceptance limit $M = 0.10$; maximum acceptable true failure rate $P = 0.20$.

Table V is used to find Z. In Table V, 0.9495 confidence gives $Z = 1.64$, and 0.9505 confidence gives $Z = 1.65$. Use $Z = 1.645$ for 0.95 confidence.

Then

$$N = 0.2(1-0.2)\left[\frac{1.645}{0.2-0.1}\right]^2$$

$$N = 43.3.$$

Partial samples aren't possible, so use $N = 44$.

Of course, since $F = MN$, $F = (0.1)(44)$, F is 4.4. Partial failures aren't allowed, so F is rounded to 4. Therefore, to verify that a unit's failure rate is less than 20 percent, there can be no more than 4 failures in 44 trials.

Comparative Testing

For *comparative testing* of two units, the normal approximation to the binomial must be used. The number of trials (N) is based on

$$N = 2P(1-P)\left[\frac{Z}{M_2-M_1}\right]^2$$

where

$$P = \frac{M_2+M_1}{2}.$$

In this case, the value of M is not an acceptance limit, but an expected failure rate. As before, there is the problem of two unknowns. In order to find N, therefore, guess (hopefully an educated guess) what the failure rate will be for one of the units being compared. Then, the failure rate of the second unit is set as a fraction, or multiple, of the first unit's failure rate. One point to keep in mind is that the more similar unit one is to unit two, the more trials will be required to determine if they are different. If the two units are expected to be identical, N must be infinite. Therefore, you must decide how much difference between units is really of interest. For example, you could state, "I expect the failure rate of unit one to be 10 percent; and I only want to know if unit two has a failure rate half as great (or twice as much)." Another important point to remember is that F for either unit (remember $F = MN$) should be at least 5, in order to obtain reasonably accurate comparisons. An example will be used to illustrate. Assume 95 percent confidence level is required. $M_1 = 0.1$ (expected failure rate for unit 1), $M_2 = 0.2$ (want to know if unit 2 is twice as bad); from Table V, $Z = 1.645$.

$$N = 2(0.15)(1-0.15)\left[\frac{1.645}{0.2-0.1}\right]^2$$

$$N = 69.$$

Therefore, if the failure rate guess is correct, it will be necessary to test each unit with 69 trials to compare them.

Of course, the obvious question is, what if the guess wasn't correct? Another way of saying this is: will the test be sufficient to achieve the confidence level required? After the test, this question can be answered by reversing the calculation process. Instead of using established confidence levels and limits to calculate test requirements, use the test results to calculate confidence levels. This is done with the following formula (which is simply a rearrangement of the previous formula).

F_1 = the measured failures of unit 1
F_2 = the measured failures of unit 2

$$Z = \frac{\dfrac{F_1 - F_2}{N}}{\sqrt{\left(\dfrac{F_1 - F_2}{N^2}\right)\left(1 - \dfrac{F_1 + F_2}{2N}\right)}}$$

The resulting value for Z is then used in Table V to find the confidence level that one unit is truly better than the other. For example, if $Z = 1.65$, then there is 95.05 percent confidence that they are somehow different. If Z is positive, the resulting confidence level is that 1 is worse than 2. If Z is negative, the confidence level is that 1 is better than 2. If the resulting confidence level is less than the goal, then it isn't certain which is better, and a new test must be done. The new guess for M_1 and M_2 would be based on the results of the first unsuccessful test.

Summary

In order to obtain accurate and repeatable results, ESD test specifications must not only detail test "hardware," but also detail statistically derived sampling requirements based on established confidence levels. For acceptance tests, the first requirement is to set a goal for the confidence level and maximum failure rates. Both the number of trials and acceptance limits must be calculated based on these goals. For comparative tests, the confidence level is also set, but the number of trials is based on estimates of the expected failure rates. After the comparative test is completed, the resulting confidence level can be established to determine if the number of trials used was indeed sufficient to accurately compare the units.

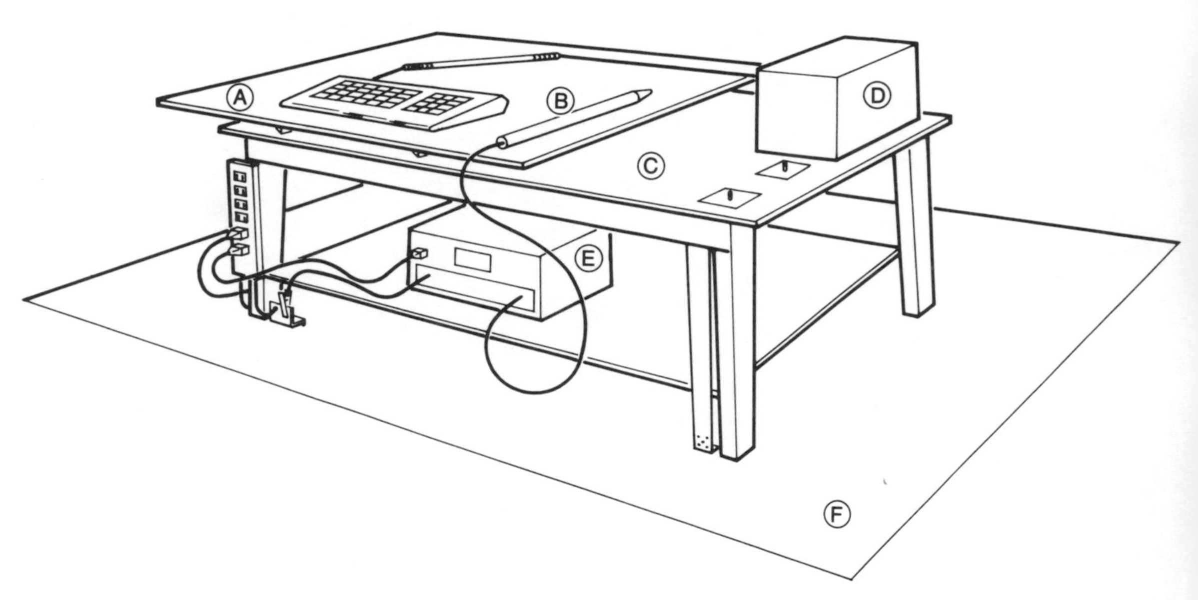

Fig. 40: *Example of an ESD test facility: A. Insulation on metal table plane; B. ESD simulator probe; C. Wooden table; D. EUT power supply and EUT end of optical link shield box; E. ESD simulator controller; F. Sheet metal ground plane with isolation transformer (unseen behind table).*

Appendix
Sample ESD Test Specification

The electrostatic discharge (ESD) test specification included in this appendix is not intended to cover all ESD test situations. This is simply an example of one implementation for testing tabletop equipment with an air discharge simulator. An example of ESD test facility is shown in Figure 40.

PURPOSE: This specification/procedure is to ensure that ESD testing is done in a uniform manner, and that the results represent a realistic measurement of the equipment's sensitivity to ESD. The physical test set-up, test methods, and test report format are covered by this procedure.

SCOPE: To be used by a Technician and Engineer who understand the operation of the ESD simulator and of the equipment under test (EUT).

RESPONSIBLE PARTY	ACTION PERFORMED
ENGINEER:	1. Determine whether a special customer, or other, ESD test is to be used, and specify on ESD test instruction form.
	2. Fill out ESD test instruction form to specify an EUT configuration as close as possible to the production configuration.
LAB TECH:	3. Use ESD test instructions provided by the Engineer to set up the equipment to be tested.
	4. Place the EUT so that it is at least four inches from the edge of the ground plane.
	5. Loop interconnect cables so they form the largest possible loops.
	6. Verify that the EUT and cables do not short to the ground plane or other equipment.
	7. Turn on the EUT power supply, and verify the EUT is operational. Set any EUT indicators on.

8. Fill out the required data on the top of the test data form for the indirect version of the appropriate test. This includes all items except the actual test data (see test form below).

9. Verify the round tip is installed on the ESD simulator probe.

10. Verify the ESD simulator is properly connected to the earth ground plane on the floor, turn voltage control to 0 V, and turn the simulator on.

11. Set the ESD simulator voltage to the lowest test voltage.

12. Move the tip of the probe rapidly toward the center front edge of the metal table plane until it discharges. Then withdraw it until the simulator voltage meter indicates it has recharged. Keep the probe perpendicular to the edge of the table during discharge. Use a consistent approach speed.

13. Record any EUT failures using the codes listed on the test data form.

14. Verify that the EUT is still operational. If not, restore it to the same state as existed prior to the discharge.

15. Repeat Steps 12 through 15 until the proper sample size is attained, or the acceptance limit is exceeded (whichever comes first.) If the acceptance limit is exceeded, go to 30.

16. Increase the voltage to the next level specified for the test, and repeat Steps 12 through 15.

17. Keep repeating these steps through the maximum indirect test voltage, applying the proper number of discharges per voltage level in each case.

18. Set simulator voltage control to 0 V. Turn the ESD simulator off and discharge the tip to chassis ground.

19. Fill out the required data on the test form for the direct discharge version of the appropriate test.

20. Turn on the ESD simulator and adjust it to the starting voltage.

21. Bring the probe tip toward the EUT discharge point, keeping it perpendicular to the EUT surface; then withdraw after discharge to allow recharging.

22. Record the EUT failures as before.
23. Verify that the EUT is still operational. If not, restore it to the same state as existed prior to the discharge.
24. Repeat Steps 21 through 23, applying the appropriate number of discharges per level according to the test and voltage level chosen.
25. Continue Steps 21 through 24 through the maximum voltage. (If a destructive failure occurs, or the EUT fails to meet the acceptance limit at any of the discharge voltage levels, then skip to Step 30.)
26. Turn off the ESD simulator and discharge the tip.
27. Turn on the ESD simulator and adjust to 25 kV. Discharge for a total of 3000 hits. Ignore nondestructive failures.
28. If the EUT is expected to be handled frequently while disconnected, a second destruction test (Step 27) should be done with the unit disconnected.
29. Verify that the EUT is still operational after the 3000 discharges. (It may be necessary to turn off the EUT power, and then turn it back on.)
30. Adjust the simulator voltage to 0 V. Turn off the static simulator and discharge the probe tip to chassis ground.
31. Disconnect the EUT and remove it from the test station.
32. Give the results of the test to the Engineer.

ENGINEER:

33. Review the test results. If the EUT passes the test, go to Step 36.
34. Analyze firmware, software, and hardware to determine possible fixes if the EUT fails to meet requirements.
35. Rework the EUT to incorporate ESD fixes and then return it to the Lab Tech for retest. Go back to Step 3.
36. File passing test results.

The failure rate level, which can be achieved, is heavily dependent on limitations inherent within the product design. Therefore, three sets of acceptance limits have been developed for three generic sets of design constraints.

STANDARD ACCEPTANCE TEST NUMBER 1 FOR ESD

When the EUT has I/O that is edge triggered and provides no parity or frame error checking, then the following limits must be met (for both direct and indirect discharges). The chassis of the EUT must be connected to chassis ground.

VOLTAGE LEVEL	NUMBER OF DISCHARGES PER LEVEL	REJECT IF THE NUMBER OF OCCURRENCES EXCEED THE FOLLOWING:		
		RECOVERABLE	NON-RECOVERABLE	DESTRUCTION
5 kV	600	24	0	—
10 kV	300	21	0	—
12.5 kV	45	9	0	—
15 kV	45	9	0	—
*25 kV	3000	—	—	0

THE ABOVE LIMITS ARE TO ENSURE THAT THE TRUE UNIT FAILURE RATE IS BETTER THAN THE FOLLOWING ESD STD**

5 kV	—	6%	0.5%	—
10 kV	—	10%	1%	—
12.5 kV	—	30%	7%	—
15 kV	—	30%	7%	—
25 kV	—	—	—	0.1%

* Destruction testing is only done with direct discharge to EUT at 25 kV.
** The above test results in 95 percent confidence that the unit failure rate is better than ESD standards.

STANDARD ACCEPTANCE TEST NUMBER 2 FOR ESD

When the EUT has I/O that is not edge triggered, and that does both parity and frame checking, then the following limits must be met (for both direct and indirect discharges). The chassis of the EUT must be connected to chassis ground.

VOLTAGE LEVEL	NUMBER OF DISCHARGES PER LEVEL	REJECT IF THE NUMBER OF OCCURRENCES EXCEED THE FOLLOWING:		
		RECOVERABLE	NON-RECOVERABLE	DESTRUCTION
5 kV	600	12	0	—
10 kV	300	11	0	—

12.5 kV	45	5	0	—
15 kV	45	5	0	—
*25 kV	3000	—	—	0

THE ABOVE LIMITS ARE TO ENSURE THAT THE TRUE UNIT FAILURE RATE IS BETTER THAN THE FOLLOWING ESD STD**

5 kV	—	3.5%	0.5%	—
10 kV	—	6%	1%	—
12.5 kV	—	19%	7%	—
15 kV	—	19%	7%	—
25 kV	—	—	—	0.1%

* Destruction testing is only done with direct discharge to EUT at 25 kV.
** The above test results in 95 percent confidence that the unit failure rate is better than ESD standards.

STANDARD ACCEPTANCE TEST NUMBER 3 FOR ESD

If the EUT has enclosure and cable shielding, and the I/O has no edge triggering, and has parity and frame checking, then the following limits must be met for both direct and indirect discharges.

VOLTAGE LEVEL	NUMBER OF DISCHARGES PER LEVEL	REJECT IF THE NUMBER OF OCCURRENCES EXCEED THE FOLLOWING:		
		RECOVERABLE	NON-RECOVERABLE	DESTRUCTION
5 kV	600	2	0	—
10 kV	300	2	0	—
12.5 kV	45	1	0	—
15 kV	45	1	0	—
*25 kV	3000	—	—	0

THE ABOVE LIMITS ARE TO ENSURE THAT THE TRUE UNIT FAILURE RATE IS BETTER THAN THE FOLLOWING ESD STD**

5 kV	—	1%	0.5%	—
10 kV	—	2%	1%	—
12.5 kV	—	11%	7%	—

15 kV	—	11%	7%	—
25 kV	—	—	—	0.1%

* Destruction testing is only done with direct discharge to EUT at 25 kV.
** The above test results in 95 percent confidence that the unit failure rate is better than ESD standards.

ESD FAILURE DEFINITIONS

1. DESTRUCTIVE FAILURE:
   This occurs when the unit must be repaired to restore operation.* An example would be destruction of a microprocessor.
2. NON-RECOVERABLE FAILURE (LOCK-UP):
   These failures are those which require a hardware reset, but not repair, to restore operation.* An example would be a computer that is locked up, or a keyboard that has an LED stuck on that requires power to be turned off, and back on, to restore.
3. RECOVERABLE FAILURE:
   These failures require operator intervention, but not hardware reset, or repair, to restore operation.* An example would be output of a false key code, which must be deleted by pressing the delete key.

* OPERATION means conformance to the specification in the same manner and configuration as existed just prior to the ESD event. This means transient conditions, such as LED flicker, are not failures.

ESD TEST INSTRUCTIONS

EUT = Equipment Under Test

1. *Test Type*:

    ____________ Std. Acceptance Test (indicate 1, 2 or 3)
    ____________ Std. Comparison Test
    (List voltage levels) ____

    ____________ Other (attach detailed instructions or acceptable customer spec).

2. *Interface For EUT That Is Not A Stand-Alone System*:

    Test EUT using ________________________ system.

    Test EUT using ________________________ interface and optional isolation link.

    ________________________ Other (attached detailed instructions)

3. *Enclosure for EUT*:

    ____________ Test EUT using enclosure provided.

    ____________ Test EUT without enclosure.

    ____________ Test EUT without enclosure, but simulate metal chassis.

    ____________ Other (attached detailed instructions)

4. *Cable for EUT*:

    ____________ Test EUT using cable(s) provided.

    ____________ Make a shielded cable to do ESD test.

    ____________ Test EUT with unshielded cable.

    ____________ Other (attached detailed instructions)

5. *Position of EUT*:

    ____________ Test with EUT setting directly on floor ground plane.

    ____________ Test with EUT setting on floor, but insulated from ground plane by rubber backed carpet.

    ____________ Test with EUT setting directly on table ground plane.

    ____________ Test with EUT setting on table, but insulated from table ground plane by 2.5 cm thick insulating surface.

6. *Chassis Ground Connection for EUT*:

    ____________ Connect chassis ground to ________________________ on EUT.

    ____________ No special chassis ground connection required.

7. *Discharge Point on EUT*:

   Discharge to the following point on the EUT: ________________________________

   ____

   (Describe exact point. Draw diagram if necessary.)

8. *Other Special ESD Test Instructions*: ________________________________

   ____

   ____

   ____

   ____

ESD TEST DATA SHEET FOR ACCEPTANCE TEST NUMBER 1

EUT P/N ____________ REV ____________ CUSTOMER ____________ DATE ____________
PROJECT ENGINEER ____________ LAB TECH ____________ PAGE ________ OF ________
ATMOSPHERIC CONDITIONS: TEMPERATURE ________ RELATIVE HUMIDITY ________ %
SOFTWARE: PART NUMBER ______________________ REV ______________________
TESTED ON: SYSTEM ______________________ INTERFACE ______________________

DIRECT/INDIRECT TEST (Circle One)

kV	# OF HITS	ERROR CODE
5.0	600	
10.0	300	
12.5	45	
15.0	45	
25.0	3000	

TEST SUMMARY (ALLOWED ERROR LIMITS IN PARENTHESIS)

TEST VOLTAGE	RECOVERABLE	NON RECOVERABLE	DESTRUCTIVE
5 kV	(24)	(0)	—
10.0 kV	(21)	(0)	—
12.5 kV	(9)	(0)	—
15 kV	(9)	(0)	—
25 kV	—	—	0

The above test ensures the unit meets the following criteria, with 95% confidence.

TEST VOLTAGE	RECOVERABLE	NON RECOVERABLE	DESTRUCTIVE
5 kV	6%	0.5%	—
10 kV	10%	1%	—
12.5 kV	30%	7%	—
15 kV	30%	7%	—
25 kV	—	—	0.1%

FAILURE CATEGORY CODES

RECOVERABLE ERROR CODE
D1—False outputs
D2—Burst of outputs
D3—LEDs change state (out of sync)
D4—Continuous outputs
D5—Input ignored
D6—Operating mode changed
D7— ____________
D8— ____________

NON-RECOVERABLE ERROR CODES
L1—Complete Lock-up
L2—Continuous outputs (unable to stop)
L3—LED or beeper stuck
L4— ____________

HARD ERROR CODES
H1—Damaged component

H2— ____________________
H3— ____________________

COMMENTS: ____

____

____

____

ESD TEST DATA SHEET FOR ACCEPTANCE TEST NUMBER 2

EUT P/N ____________ REV ____________ CUSTOMER ____________ DATE ____________
PROJECT ENGINEER __________ LAB TECH __________ PAGE __________ OF __________
ATMOSPHERIC CONDITIONS: TEMPERATURE ________ RELATIVE HUMIDITY ________ %
SOFTWARE: PART NUMBER ______________________ REV ______________________
TESTED ON: SYSTEM ______________________ INTERFACE ______________________

DIRECT/INDIRECT TEST (Circle One)

kV	# OF HITS	ERROR CODE
5.0	600	
10.0	300	
12.5	45	
15.0	45	
25.0	3000	

TEST SUMMARY (ALLOWED ERROR LIMITS IN PARENTHESIS)

TEST VOLTAGE	RECOVERABLE	NON RECOVERABLE	DESTRUCTIVE
5 kV	(12)	(0)	—
10.0 kV	(11)	(0)	—
12.5 kV	(5)	(0)	—
15 kV	(5)	(0)	—
25 kV	—	—	(0)

The above test ensures the unit meets the following criteria, with 95% confidence.

TEST VOLTAGE	RECOVERABLE	NON RECOVERABLE	DESTRUCTIVE
5 kV	3.5%	0.5%	—
10 kV	6%	1%	—
12.5 kV	19%	7%	—
15 kV	19%	7%	—
25 kV	—	—	0.1%

FAILURE CATEGORY CODES

RECOVERABLE ERROR CODE

D1—False outputs
D2—Burst of outputs
D3—LEDs change state (out of sync)
D4—Continuous outputs (able to stop)
D5—Input ignored
D6—Operating mode changed
D7— ____________
D8— ____________

NON-RECOVERABLE ERROR CODES

L1—Complete Lock-up
L2—Continuous outputs (unable to stop)
L3—LED or beeper stuck
L4— ____________

HARD ERROR CODES
H1—Damaged component
H2— ____________________
H3— ____________________

COMMENTS: ____

____

____

____

ESD TEST DATA SHEET FOR ACCEPTANCE TEST NUMBER 3

EUT P/N ____________ REV ____________ CUSTOMER ____________ DATE ____________
PROJECT ENGINEER __________ LAB TECH __________ PAGE __________ OF __________
ATMOSPHERIC CONDITIONS: TEMPERATURE ________ RELATIVE HUMIDITY ________ %
SOFTWARE: PART NUMBER ________________________ REV ________________
TESTED ON: SYSTEM ______________________ INTERFACE ________________

DIRECT/INDIRECT TEST (Circle One)

kV	# OF HITS	ERROR CODE
5.0	600	
10.0	300	
12.5	45	
15.0	45	
25.0	3000	

TEST SUMMARY (ALLOWED ERROR LIMITS IN PARENTHESIS)

TEST VOLTAGE	RECOVERABLE	NON RECOVERABLE	DESTRUCTIVE
5 kV	(2)	(0)	—
10.0 kV	(2)	(0)	—
12.5 kV	(1)	(0)	—
15 kV	(1)	(0)	—
25 kV	—	—	(0)

The above test ensures the unit meets the following criteria, with 95% confidence.

TEST VOLTAGE	RECOVERABLE	NON RECOVERABLE	DESTRUCTIVE
5 kV	1%	0.5%	—
10 kV	2%	1%	—
12.5 kV	11%	7%	—
15 kV	11%	7%	—
25 kV	—	—	0.1%

FAILURE CATEGORY CODES

RECOVERABLE ERROR CODE
D1—False outputs
D2—Burst of outputs
D3—LEDs change state (out of sync)
D4—Continuous outputs (able to stop)
D5—Input ignored
D6—Operating mode changed
D7— ________________
D8— ________________

NON-RECOVERABLE ERROR CODES
L1—Complete Lock-up
L2—Continuous output (unable to stop)
L3—LED or beeper stuck
L4— ________________
HARD ERROR CODES

H1—Damaged component
H2— ____________________
H3— ____________________

COMMENTS: ____

____

____

____

STANDARD COMPARATIVE TEST CRITERIA FOR ESD

A standard comparison test is only done for those voltage levels specifically indicated by the person requesting the test. For example, a unit which is already in production may be compared against a newly modified version of the same unit. However, it may only be desired to compare the units at those voltages where the modified unit fails acceptance testing. The purpose of this example comparison would be to verify that the modified unit either passes present acceptance levels or is at least no worse than the previous design. (Totally new designs must always pass present acceptance test levels.)

The comparison test consists of both direct and indirect ESD tests. The number of discharges for each voltage level, for the first EUT, is either 1600, or the number of discharges necessary to obtain 10 failures, whichever is less. The second EUT will use the same number of discharges at each voltage level as first EUT.

The comparison decision is obtained from the following chart:

If the lower failure rate is less than:	Then the higher failure rate must be greater than:
0.5%	1%
1%	2%
2%	4%
5%	10%
10%	20%

If the higher failure rate is greater than the value given above, then we have at least 95 percent confidence that the EUT with the lower failure rate truly is better than the EUT with the higher failure rate.

ESD TEST DATA SHEET FOR COMPARISON TESTING

EUT P/N ____________ REV ____________ CUSTOMER ____________ DATE ____________
PROJECT ENGINEER __________ LAB TECH __________ PAGE __________ OF __________
ATMOSPHERIC CONDITIONS: TEMPERATURE ________ RELATIVE HUMIDITY ________ %
SOFTWARE: PART NUMBER ________________________ REV ____________________
TESTED ON: SYSTEM ________________________ INTERFACE ____________________

DIRECT/INDIRECT TEST (Circle One)

kV (Circle)	# OF HITS (Max. 1600)	ERROR CODE (If there are less than 1600 discharges, 10 errors must be shown.)
5.0		
10.0		
12.5		
15.0		

If the lower failure rate is less than	Then the higher failure rate must be greater than
0.5%	1%
1%	2%
2%	4%
5%	10%
10%	20%

If these conditions are met, there will be 95% confidence that the lower EUT is really better than the higher EUT.

COMMENTS: ________________________________

____

____

____

____

____

____

FAILURE CATEGORY CODES

RECOVERABLE ERROR CODES

D1—False outputs
D2—Burst of outputs
D3—LED's change state (out of sync)
D4—Continuous outputs (able to stop)
D5—Input ignored
D6—Operating mode change
D7—____________________
D8—____________________

NON-RECOVERABLE ERROR CODES

L1—Complete lock-up
L2—Continuous outputs (unable to stop)
L3—LED or beeper stuck
L4—____________________

HARD ERRORS

H1—Damaged component
H2—____________________
H3—____________________

Bibliography

Achenback, B., ''Particle Size Compaction and its Effect on Conductive Coatings,'' *EMC Tech.,* vol. 4, pp. 37–40, Oct.–Dec. 1985.

Anderson, D. C., ''Such Stuff as Sparks are Made Of,'' *Electron. Rep.,* pp. 17–20, Oct. 1985.

Arpin, Kevin J., ''EMI Control in Light-Duty Cables via Shields and Connectors,'' *EMC Tech.,* Sept.–Oct. 1986, pp. 19–25.

Banks, M., ''Non-Insulating Thermoplastic Composites,'' *ITEM,* pp. 273–274, 1986.

Benson, P., ''Plastic Housing Design and EMI Shielding,'' *ITEM,* pp. 282–292, 1985.

Berbeco, G. R., ''Static Elimination in the Workplace,'' *ITEM,* pp. 80–82, 1985.

Bhar, T., ''Electrostatic Discharge Failure Mechanisms of Semiconductor Devices,'' *ITEM,* pp. 276–282, 1986.

Bogar, J., and Sterling, D., ''Minimize EMI in Your Computer Design,'' *Electrostatic Products,* pp. 105–111, Jan. 1984.

Bogar, J., and Vander Heyden, E., ''Interconnection and Circuit Packaging for Electro-Magnetic Compatibility,'' AMP Inc. Rep., Harrisburg, PA, 1982.

Byrne, W., ''Development of Design and Test Procedures to Meet Electrostatic Discharge (ESD),'' presented at MIDCON Convention, Dallas, TX, Nov. 30–Dec. 2, 1982

Byrne, W., ''The Meaning of Electrostatic Discharge (ESD) in Relation to Human Body Characteristics and Electronic Equipment,'' presented at *IEEE Int. Symp. Electromagnetic Compatibility,* Washington DC, Aug. 23–25, 1983.

Carlson, E., and Tzeng, W., ''Galvanic Corrosion Measurements on EMI-Gasketed Joints,'' *ITEM,* pp. 312–314, 1985.

Chubb, J. ''Instruments and Approaches for Static Measurements,'' *ITEM,* pp. 296–324, 1986.

Clarke, R., ''Conductive EMI/RFI Shielding Materials Answer Challenge Posed by FCC Regulation,'' *EDN,* pp. 83–92, Jan. 1985.

Dash, G., ''Designing to Avoid Static-ESD: Testing of Digital Devices,'' *ITEM,* pp. 96–110, 1985.

Dash, G., ''Thorough ESD Testing Prevents Digital Device Field Failures,'' *EDN,* pp. 213–220, Aug. 1984.

Daout, B., and Ryser, H., ''The Reproducibility of the Rising Slope in ESD Testing,'' in *Proc. IEEE Int. Symp. EMC,* San Diego, CA, Sept. 16–18, 1986.

Daout, B., and Ryser, H., ''Fast Discharge Mode in ESD Testing,'' in *Proc. 6th EMC Symp.,* Zurich, Switzerland, 1985, pp. 41–46.

Della Torre, E., and Longo, C., *The Electromagnetic Field.* Boston, MA: Allyn and Bacon, Inc., 1969.

Dietz, P., ''Selective Electroplated Deposits for the Control of EMI,'' *ITEM,* pp. 172–174, 1986.

Don White Consultants, ''Introduction to EMI/EMC,'' Seminar, Los Angeles, CA, Dec. 5–7, 1979.

Electronic Industries Association, ''Designers Guide on Electromagnetic Compatibility, Bonding of Electronic Equipment,'' EMC Bulletin No. 5, Feb. 1964.

Electronic Industries Association, ''Designers Guide on Electromagnetic Compatibility, Cabling of Electronic Equipment,'' EMC Bulletin No. 8, Mar. 1965.

Electronic Industries Association, ''Designers Guide on Electromagnetic Compatibility, Enclosures for Electronic Equipment,'' EMC Bulletin No. 7, 1966.

Electronic Industries Association, ''Designers Guide on Electromagnetic Compatibility, Grounding of Electronic Equipment,'' EMC Bulletin No. 6, Dec. 1967.

Genova, A., ''Testing the Shielding Effectiveness of Conductive Paints,'' *Plastic Design and Processing,* Oct.–Nov., 1982.

Georgopoulos, C., ''EMI Control in the Installation and Interconnection of Digital Equipment,'' *EMC Tech.,* vol. 5, no. 2, pp. 55–63, Mar.–Apr. 1986.

Gerke, D., and Kimmel, W., ''Design Noise Tolerance into Microprocessor Systems,'' *EMC Tech.,* vol. 5, no. 2, pp. 45–52, Mar.–Apr. 1986.

Gindrup, W., and Vinson, R., ''A Logical Approach to EMI Shielding Materials,'' *ITEM,* pp. 184–352, 1986.

Halprin, S., ''Static Control Specification Development for the Electronic Industry,'' *ITEM,* pp. 194–200, 1984.

Hollander, D., ''The Hidden Dangers of Electrostatic Discharge (ESD),'' *QST,* Mar. 1987, pp. 38–40.

Honda, M., and Takeyoshi, K., ''ESD Characteristics and Their Effects on Computers,'' *Electromagnetic Commun. J.*: vol. 83, no. 209, pp. 25–30, Dec. 1983; vol. 83, no. 292, pp. 13–17, Mar. 1984; vol. 85, no. 86, pp. 39–42, July 1985; vol. 86, no. 107, pp. 17–22, July 1986.

Hopkins, M., ''Testing Equipment and Circuits for ESD Sensitivity,'' presented at Electrical Overstress Exposition, Anaheim, CA, Jan. 21–23, 1986.

Huntsman, J., ''Electric Fields, Static Damage, and Shielding,'' 3M Company, St. Paul, MN.

Huntsman, J., and Yenni, D., "Charge Drainage vs. Voltage Suppression by Static Control Table Tops," *Evaluation Eng.*, Mar. 1982.

Huntsman, J., Yenni, D., and Mueller, G., "Fundamental Requirements for Static Protective Containers," presented at NEPCOM/West, Anaheim, CA, 1980.

Hyatt, H., EMI/RFI Seminar Notes, Experimental Physics Corp., Hayward, CA, Feb. 1982.

Hyatt, H., Calvin, H., and Mellberg, H., "Measurement of Fast Transients and Application to Human ESD," presented at EOS/ESD Symp., San Diego, CA, Sept. 1980.

Hyatt, H., Calvin, H., and Mellberg, H., "A Closer Look at the Human ESD Event," presented at EOS/ESD Symp., Las Vegas, NV, Sept. 1981.

Intel Corp., "Designing Microcontroller Systems for Electrically Noisy Environments," Rep. AFN-02131A, Santa Clara, CA, 1986.

International Electrotechnical Commission, "Electromagnetic Compatibility for Industrial Process Measurement and Control Equipment," Publication 801-2, 1984.

KeyTek Instruments Corp., *Electrostatic Discharge (ESD).* Wilmington, MA, 1986.

King, W., and Reynolds, D., "Personnel Electrostatic Discharge: Impulse Waveforms Resulting From ESD of Humans Directly and Through Small Hand-Held Metallic Objects Intervening in the Discharge Path," in *Proc. IEEE Int. Symp. EMC,* 1981, pp. 577–590.

King W., and Reynolds, D., "Personnel Electrostatic Discharge: Impulse Waveforms Resulting From ESD of Humans Through Metallic-Mobile Furnishings Intervening in the Discharge Path," in *Proc. IEEE Int. Symp. EMC,* Sept. 8–10, 1982, pp. 220–225.

King, W., "Dynamic Waveform Characteristics of Personnel Electrostatic Discharge," in *Proc. EOS/ESD Symp.*, Denver, CO, Sept. 24–27, 1979, pp. 78–87.

Kolyer, J., and Cullop, D., "Getting ESD Floor-Finish Tests Rolling Toward Standardization," *EOS/ESD Tech.*, pp. 20–23, Apr. 1987.

Kunz, H., "Electrostatic Charging and Simulation of the Discharging Process," *EMC Tech.,* pp. 30–36, Apr. 1982.

Ling, J., "EMI Shielding: Selection of Materials for Conductive Coatings," *ITEM,* pp. 186–188, 1986.

Lutz, M., Frey, O., and Taylor, W., "Testing for Human ESD Control," *ITEM,* pp. 284–325, 1986.

Laine, A., "Concepts of Shielding," *ITEM,* pp. 322–328, 1985.

Maloney, T., "Contact Injection: A Major Cause of ESD Failure in Integrated Circuits," presented at *EOS/ESD Symp.,* Las Vegas, NV, Sept. 23–25, 1986.

Maloney, T., and Khurana, N., "Transmission Line Pulsing Techniques for Circuit Modeling of ESD Phenomena," presented at *EOS/ESD Symp.,*1985.

Maloney, T., and Khurana, N., "ESD on CHMOS Devices–Equivalent Circuits, Physical Models of Failure Mechanisms," in *Proc. IEEE/IRPS* 1985.

Mardiguian, M., "The ESD Simulator, an EMI Engineer's Multi-Faceted Friend," *EMC Tech.,* pp. 55–61, Jan.–Mar. 1985.

Mardiguian, M., *Electrostatic Discharge.* Gainsville, VA: Interference Control Technologies, Inc. 1986.

Mardiguian, M., and White, D., "Electrostatic Discharge, What it is and How to Control it," *Electronic Products,* pp. 111–115, Sept. 1983.

McAuliffe, T., "Conductive Adhesive Foil Shielding Tape," *EMC Tech.,* pp. 53–54, Jan.–Feb. 1986.

Miller, I., and Freund, J., *Probability and Statistics for Engineers.* Englewood Cliffs, NJ: Prentice-Hall, 1965.

Minton, R., Bastenbeck, E., and Shawhan, G., "Multilayer Electroless Coatings for EMI Shielding," *ITEM,* pp. 176–184, 1986.

Mooney, J., "Plastics EMI Shielding: The Evolving State of the Art," *EMC Tech.,* vol. 4, no. 4, pp. 19–28, Oct.–Dec. 1973.

Moore, A. D., *Electrostatic and Its Application.* New York, NY: Wiley & Sons, 1973.

Morrison, R., *Grounding and Shielding Techniques in Instrumentation,* 2nd Ed. New York, NY: Wiley & Sons, 1977.

Moss, R., "Exploding the Humidity Half-Truth and Other Dangerous Myths," *EOS/ESD Tech.,* pp. 10–19, Apr. 1987.

Parker, C., Tolen, B., and Parker, R., "Prayer Beads Solve Many of Your EMI Problems," *EMC Tech.,* pp. 39–70, Apr.–June 1985.

Rich, A., "Understanding EMI-Type Noise," *Electronics Test,* May, 1983.

Richman, P., "An ESD Circuit Model with Initial Spikes to Duplicate Discharges from Hands with Metal Objects," *EMC Tech.,* pp. 53–59, Apr.–June 1985.

Richman, P., "Calibrating ESD Simulators: Measurements Versus Computer Models," *ITEM,* 1985.

Richman, P., "Comparing Computer Models to Measured ESD Events," in *Proc. Electrical Overstress Exposition,* Boston, MA, Apr. 9–11, 1985, pp. 110–114.

Richman, P., "Computer Modeling the Effects of Oscilloscope Bandwidth on ESD Waveforms, Including Arc Oscillations," in *Proc. IEEE Int. Symp. EMC,* Wakefield, MA, Aug. 20–22, 1985, pp. 238–245.

Richman, P., "Classification of ESD Hand/Metal Current Waves Versus Approach Speed Voltage, Electrode Geometry, and Humidity," in *Proc. IEEE Int. Symp. EMC,* San Diego, CA, Sept. 16–18, 1986, pp. 451–458.

Richman, P., "The Effects of Hand-Associated Metal Object Geometry and Hand-To-Object Coupling Impedance on ESD Current Waves," *EMC Symp.,* Zurich, Switzerland, 1986.

Richman, P., and Tasker, A., "ESD Testing: The Interface Between

Simulator and Equipment Under Test,'' in *Proc. 6th EMC Symp.,* Zurich, Switzerland, 1985, pp. 25–30.

Scarlett, J., ''PC Design Basics: Power Distribution and Track Routing,'' *Printed Circ. Design,* pp. 3–7, July 1985.

Schaffner EMC Inc., *Electromagnetic Compatibility.* Union, NJ, June 1985.

Sears, F., and Zemansky, M., *University Physics,* third Ed. New York, NY: Addison-Wesley, 1964.

Sigmond, R., ''The Residual Streamer Channel: Return Strokes and Secondary Streamers,'' *J. Appl. Phys.,* vol. 56, no. 5, pp. 1355–1369, Sept. 1984.

Simpson, R. W., ''Fabrication of Flexible Foil Laminates for EMI Shielding Applications,'' *ITEM,* pp. 330–332, 1985.

Tasker, A., ''ESD Discharge Waveform Measurement, The First Step in Human ESD Simulation,'' in *Proc. IEEE Int. Symp. EMC,* Wakefield, MA, Aug. 20–22, 1985, pp. 246–250.

Tecknit Corp., *Design Guide to the Selection and Application of EMI Shielding Materials.* Cranford, NJ: Tecknit, 1982.

Terman, F., *Electronic and Radio Engineering,* fourth Ed. New York, NY: McGraw-Hill Co., 1955.

Thompson, D., ''Electroless Shielding of Plastic Electronic Enclosures,'' *EMC Tech.,* vol. 4, no. 4, pp. 43–48, Oct.–Dec. 1985.

Trompeter, E., ''Wiring and Cabling of Electronic Systems,'' *ITEM,* pp. 168–172, 1985.

Van Doren, T., *Grounding and Shielding Electronic Instrumentation.* Rolla, MO: University of Missouri, 1984.

Violette, J. L. N., and Violette, M. F., ''ESD Case History—Immunizing a Desktop Business Machine,'' *EMC Tech.,* pp. 55–60, May–June, 1986.

White, D., *EMC Control Methods and Techniques.* Germantown, MD: Don White Consultants, 1973.

White, D., ''Eliminate the Guesswork from Shielding Enclosure Design—The Impact of Apertures,'' *EMC Tech.,* pp. 29–37, Apr.–June 1985.

White, R., ''Installation Practices for Transient Protection Devices,'' *EMC Tech.,* pp. 33–39, May–June 1986.

Wong, W. S., ''ESD Design Maturity Test for a Desktop Digital System,'' *Evaluation Eng.,* pp. 104–112, Oct. 1984.

Woodburn, G., and Miller, J., ''ESD Protection with Plastic Composites,'' *ITEM,* pp. 285–290, 1986.

Yenni, D., ''Basic Electrical Considerations in the Design of Static-Safe Work Environment,'' presented at NEPCON/West, Anaheim, CA, 1979.

Index

About the Author

Warren Boxleitner (M'71, M'87) is the Director of Engineering at KeyTek Instrument Corporation, a manufacturer of ESD and surge simulation equipment. Previously he was employed by Key Tronic Corporation, which manufactures computer input devices. There he held various positions including that of International Sales Manager, Senior Engineer, and Technical Services Group Leader. Prior to that he served four years as a Communications Systems Officer in the U. S. Air Force. He holds a B.S.E.E. from the University of Idaho.

Warren is presently secretary of Working Group 3.6.8 within the IEEE Surge Protective Devices Committee of the Power Engineering Society. This Working Group is developing a guide to characterize the actual ESD event. Warren is also a member of Subcommittee 1 within the American National Standards Institute's ASC C63 Techniques and Developments Committee. This group is developing an ESD testing standard. In addition, he also serves as a member of an EOS/ESD Association committee which is developing an ESD test standard for components.